How I changed my mind about

EVOLUTION

Evangelicals Reflect on FAITH and SCIENCE

EDITED BY

Kathryn Applegate
& J. B. Stump

D0061697

IVP Academic
An imprint of InterVarsity Press
Downers Grove, Illinois

InterVarsity Press
P.O. Box 1400, Downers Grove, IL 60515-1426
ivpress.com
email@ivpress.com

InterVarsity Press® is the book-publishing division of InterVarsity Christian Fellowship/USA®, a movement of students and faculty active on campus at hundreds of universities, colleges and schools of nursing in the United States of America, and a member movement of the International Fellowship of Evangelical Students. For information about local and regional activities, visit intervarsity.org.

All Scripture quotations, unless otherwise indicated, are taken from THE HOLY BIBLE, NEW INTERNATIONAL VERSION®, NIV® Copyright ©1973, 1978, 1984, 2011 by Biblica, Inc.™ Used by permission. All rights reserved worldwide.

Cover design: Cindy Kiple
Interior design: Beth McGill
Images: Imagezoo / Superstock / Glow Images

ISBN 978-0-8308-5290-1 (print)
ISBN 978-0-8308-9963-0 (digital)

Printed in the United States of America ∞

Library of Congress Cataloging-in-Publication Data

Names: Applegate, Kathryn, 1982- editor.

Title: How I changed my mind about evolution : evangelicals reflect on faith and science / edited by Kathryn Applegate and J. B. Stump.

Description: Downers Grove : InterVarsity Press, 2016. | Includes bibliographical references.

Identifiers: LCCN 2016007931 (print) | LCCN 2016009334 (ebook) | ISBN 9780830852901 (pbk. : alk. paper) | ISBN 9780830899630 (eBook)

Subjects: LCSH: Evolution (Biology)—Religious aspects—Christianity. | Religion and science. | Evangelicalism.

Classification: LCC BL263 .H69 2016 (print) | LCC BL263 (ebook) | DDC 231.7/652—dc23

LC record available at http://lccn.loc.gov/2016007931

P   20   19   18   17   16   15   14   13   12   11   10   9   8   7   6   5   4   3   2   1

Y   33   32   31   30   29   28   27   26   25   24   23   22   21   20   19   18   17   16

"This collection of firsthand experiences is important for showing that firm belief in the truth-telling character of Scripture can support, rather than undermine, the best scientific investigations. It also provides more solid evidence for the good that BioLogos is doing to transform science and religion from a war zone to an instructive conversation."

Mark Noll, Francis A. McAnaney Professor of History, University of Notre Dame

"This book captures the convictions and stories of an array of Christians whose scholarship, reflection, faith and community have brought them to affirm God's use of evolution in the processes of creation. Consider their stories. Ponder their convictions. May your journey too be one that fully opens to worshipful wonder and to scientific discovery."

Mark Labberton, president, professor of preaching, Fuller Theological Seminary, author of *Called*

"While very few people seem to care about science as a philosophical construct, many people care about how scientific findings impact their understanding of life, love, meaning and faith. No scientific concept draws us into these contemplations more than evolution, which leads us to ask big questions about our nature and God's. This book should be embraced as a treasure. In it you will find unique minds wrestling with how we got here and what our evolutionary history has to do with God, the Bible and the depth of our lived experience. I promise this book will challenge you on nearly every page as you discover new insight after new insight."

Andrew Root, Olson Baalson Associate Professor of Youth and Family Ministry, Luther Seminary, principal leader, Science for Youth Ministry

"Atheists often cite religious opposition to evolution as a reason for their unbelief. This wonderful collection of essays by Bible-believing Christians demonstrates how unnecessary it is to oppose evolution in the name of faith. What is striking about the authors in this volume is the sheer range and diversity of their own spiritual journeys in coming to terms with science. It is my prayer that evolution might cease to be seen as a threat to faith on the part of some Christians rather than as an integral aspect of God's created order for which the Christian can rightly give praise."

Denis R. Alexander, emeritus director, The Faraday Institute for Science and Religion

"If we want to converse with someone, we must first be willing to listen, to understand. And these stories are easy to listen to—so well written, so personally engaging, so reflective of committed faith. You will find yourself liking the contributors as you get a glimpse into their thought processes. As someone who studies and ministers in the area of faith and science, I commend the authors and editors for this nicely done book."

C. John "Jack" Collins, professor of Old Testament, Covenant Theological Seminary

"The conflict between Christian faith and science is long-standing and significant, and no issue has been more central to this perceived tension than evolution. In this welcome collection of essays, leading Christian thinkers explain their reasons for affirming evolution while remaining committed to their faith. It is to be hoped that this volume will find a wide readership, especially among those who struggle to relate their faith to the consensual canons of science."

John R. Franke, theologian in residence, Second Presbyterian Church, Indianapolis

This book is lovingly dedicated to our children—

Kathryn's Lucy and Josiah,

and Jim's Casey, Trevor and Connor—

in the hope that their generation will never be

forced to choose between faith and science.

Contents

Foreword

Deborah Haarsma
President, BioLogos

■ · ■ · ■

IF YOU PICKED UP THIS BOOK, YOU ARE PROBABLY curious about how science fits with Christianity. Maybe you wonder how an evangelical Christian could possibly consider an "atheistic" idea like evolution. Or maybe you are skeptical that a successful scientist could accept the "superstitious" ideas of Christianity.

At BioLogos, we hear these questions every day. On one level, these questions are intellectual—answers can be found by digging into the scientific evidence, the theological arguments and the Bible itself. At biologos.org we provide plenty of resources to help you do just that.

But on another level, these questions are personal. They engage the heart and soul as well as the mind, going beyond ideas to impact relationships within families and communities. Answers won't be found solely in intellectual arguments, and sometimes piling on more evidence doesn't help. In this volume we invite you to explore these questions by hearing the personal stories of twenty-five people who have walked this road before.

In these pages, you'll read about the real-life experiences of scientists, theologians and others as they encounter evolution and Christian faith. Their stories tell of confusion and conflict, including some sharp critiques of anti-evolutionary views from those who were casualties of such conflicts. But these pages also tell of repentance and renewal, as people discover Jesus Christ as their personal Savior *and* as Lord of all creation. These stories tell of hope and celebration, praising God for the wonders of his creation and learning how to be the body of Christ together despite our differences.

Jesus said, "You shall love the Lord your God with all your heart and with all your soul and with all your strength and with all your mind, and your neighbor as yourself" (Luke 10:27). In these stories, we see how the contributors have come to love God with their minds—including all the evidence of modern science—as well as their hearts and their souls. As we seek to engage these questions together, with humility and with confidence in the Spirit's leading, new ways of understanding will emerge that are faithful to God's Word *and* consistent with his works as revealed in creation.

Whatever your views on evolution and Christianity, I hope these stories will help you step into the lives of others, to better understand their struggles, their motivations and their faith. By hearing each other's stories, we will understand better how we can love our neighbors as ourselves, as Jesus commanded. The church doesn't need to be unified in its view of creation, but it must be unified in Christ. For "by this everyone will know that you are my disciples, if you love one another" (John 13:35).

Acknowledgments

■ · ■ · ■

THE CONCEPT FOR THIS BOOK ORIGINATED IN 2011, when philosopher and theologian Thomas J. Oord proposed it to the leaders of BioLogos and a publisher, Mark Russell. The original plan was to collect between fifty and sixty short essays by evangelicals who accept evolution, something Tom later accomplished (entirely within his own denomination!) in his book with Sherri B. Walker, *Nazarenes Exploring Evolution* (SacraSage Press, 2013). Later, and in consultation with InterVarsity Press (IVP), we decided to feature half the number of essays but allow the authors more space to share their often-complex journeys toward reconciliation between the Bible and evolutionary science. We are indebted to Tom and Mark, as well as to then-president of BioLogos, Darrel R. Falk, for their early work and ongoing support of the project.

Almost all of the chapters are brand-new essays written specifically for this book. A few of them have previously appeared on the BioLogos blog. The selections by Francis Collins and N. T. Wright are used with permission from their books, and the essay by John Ortberg is adapted with permission from one of his sermons. All of the contributors have been a pleasure to work with. Reading and rereading their stories has enriched our own. We pray the courage they've shown in sharing their stories will embolden others to do the same.

We're grateful to IVP and particularly to Andy Le Peau for shepherding this project through to completion. We're pleased to present this book as the first in a new series at IVP called BioLogos Books on Science and Christianity (see the description in the back of this book for more information). Partnering with a premier Christian publisher like IVP is a great privilege for BioLogos, and we look forward to the appearance of many new and stimulating titles in the future.

BioLogos is a tremendously rewarding place to work. President Deborah Haarsma fosters an environment in which we have important projects to work on and the resources to get them done. The rest of the staff consistently contribute in ways that enhance our work and make the office an enjoyable place to be.

BioLogos would not exist if not for the vision of Francis Collins and others. This book is evidence that the work Francis began when he founded BioLogos in 2007 is still going strong. That work has helped open lines of dialogue about science within American evangelicalism, without which stories like these could not be told.

Finally, we are grateful for the support and patience of our families as we labored over this book. Indeed, it is only because they have been so supportive that we were able to meet our deadlines and keep the home fires burning at the same time. In a very real sense Kathryn's husband, Brent, and Jim's wife, Chris, should share the recognition for this project.

Introduction

Kathryn Applegate and J. B. (Jim) Stump

■ ▪ ■ ▪ ■

EVERYBODY LOVES A STORY. The genre of memoir has become increasingly popular among the reading public. Sometimes these stories are interesting because they are so different from our own. Maybe we read about someone's experience of growing up in the circus, traveling to Nepal or living for a year without the Internet. Compelling stories capture our attention and give us a glimpse of what it's like to see the world through another's eyes. Other stories interest us because they mirror some part of our own experience. They show us that we are not alone, and the best of these stories help us navigate and interpret what we have seen and felt in ways that enrich our lives.

We hope this book can serve both of these purposes. Undoubtedly, some people reading these pages are deeply suspicious of evolution. Perhaps they've seen Richard Dawkins, that ardent defender of evolution, sneer at religion and call it a "virus of the mind." Or maybe they've heard Ken Ham, a young-earth creationist with an audience of millions, warn that "evolution and millions of years"—what he summarily dismisses as "man's word"—are baseless ideas that contradict the clear message of Genesis and inevitably lead down the slippery slope to atheism, or worse, liberal Christianity. More nuanced views are often drowned out by the polarizing rhetoric at either extreme.

BioLogos represents another choice. Our mission is to invite the church and the world to see the harmony between science and biblical faith as we present an evolutionary understanding of God's creation. Some of us are believing scientists who find the weight of evidence for evolution so strong we would do injustice to God's message in creation if we didn't speak out. Others are biblical scholars and theologians—including some who argue passionately for the historicity of Adam and Eve—who see no scriptural warrant for rejecting biological evolution, even of humans. They are grieved by the way Scripture is often forced to answer twenty-first century questions that it was never intended to address. Pastors and educators in our community see firsthand the devastating impact of the false creation-or-evolution dichotomy our Christian subculture has embraced so thoroughly. They see young people encountering compelling evidence for evolution and feeling forced to choose between science and faith.

According to a recent Gallup poll, 69% of Americans who attend church weekly believe that God created humans in their present form less than ten thousand years ago. In fact, the majority of committed Christians are unaware that it's possible to accept the overwhelming scientific evidence for evolution while maintaining a vibrant faith in Jesus Christ as Lord and Savior. More than half a million people visited the BioLogos website in the last year, and we regularly hear from new readers who until recently had never met a Christian who accepts evolution. If that describes you, allow us to introduce you to some! These twenty-five first-person narratives—minimemoirs, if you will—may offer you a glimpse of how the world of science and faith looks through the eyes of devout Christians who accept the science of evolution. They come from a wide variety of backgrounds and have taken different paths to accepting evolution. The stories don't present much of the technical evidence for the truth of evolution—which can be found in many other places (for instance, at biologos. org)—but the stories collected here give overwhelming evidence for the fact that serious Christians, who love Jesus and are committed to the authority of the Bible, can also accept evolution.

For those who already identify with the evolutionary creation position represented here, we hope you will find elements of these stories that resonate with your own. We all need exemplars—people with whom we can identify who have gone before us. Several of the contributions in this book note the importance of mentors or a supportive Christian community within which ideas could be freely explored. BioLogos has become that kind of community for many people. We invite you to join the dialogue on our website, or send us your own story (stories@biologos.org).

Part of what makes for a good story is the development of the main character. Despite the intimations of the title, not everyone here describes a profound conversion from young-earth creationism or another anti-evolutionary viewpoint. (N. T. Wright, the celebrated British New Testament scholar, describes the anti-evolution sentiment in America as an exotic oddity.) However, the majority of the authors describe a kind of cognitive dissonance they experienced while working to piece together a coherent view of the world which takes account of both Christian faith and evolutionary science. This dissonance results from the pervasive cultural message that science and Christianity are at war: that they offer competing answers to the same question and that we must choose which one to trust. When we assume that either science trumps religion or religion trumps science, we're caught in a dilemma.

It doesn't take long for the reflective Christian to realize that neither science nor Christianity has *all* the answers. Science can't tell us much about Jesus Christ or the way to have a relationship with God; and you can search the Bible from Genesis to Revelation and you won't find any descriptions of DNA or quantum mechanics! Some questions are obviously scientific and some are obviously religious. The difficulty comes when both seem relevant, as in the question of humanity's origin. For cases like this, the way forward is to allow science and faith to dialogue with each other. Learn the best science. Talk to religious thinkers you trust. Give grace to everyone, remembering that our human attempts at knowing are finite and provisional.

A related theme you'll see surfacing again and again throughout these stories is the commitment that all truth is God's truth. Whether truth is found in Scripture or through careful study of the natural world (even when that study is undertaken by unbelieving scientists!), our contributors see God as the ultimate source of all truth. This gives us unshakable confidence that there will ultimately be no contradiction between science and theology. God is the author of both. Sometimes this is referred to as the "Two Books" model of revelation. Psalm 19 captures both of these: "The heavens declare the glory of God" (v. 1) and "The law of the Lord is perfect" (v. 7). They are complementary.

Finally, both sources—God's Word and his world—drive our contributors to wonder and worship. We believe that God has given us minds and curiosity. Applying these through scientific endeavors can be an expression of love for God. Far from eliminating any sense of awe at creation, a scientific understanding of how the natural world works brings an even greater appreciation for its Creator. It is not uncommon at all for the believing scientist to report being drawn closer to God while working in the field or laboratory. Humility, wonder and worship are common themes throughout this book.

We hope that as you read these stories, you too will be drawn closer to God. We hope you'll better understand his love and provision for you and for his world, and see how he has been at work in the lives of the men and women who have so graciously shared their stories in this book.

From Culture Wars to Common Witness

A Pilgrimage on Faith and Science

James K. A. Smith

James K. A. Smith *is professor of philosophy at Calvin College and a senior fellow of The Colossian Forum. His most recent book is* Who's Afraid of Relativism? *(Baker Academic, 2014). He and his wife, Deanna, have four children and are committed urban gardeners.*

■ ▪ ■ ▪ ■

STRANGELY (AND SADLY) ENOUGH, it was Christians who taught me how to fight. Since I was not raised in a Christian home, I didn't receive the standard evangelical formation in the faith (Christian schooling, youth group, summer camps concluding with heartfelt renditions of Michael W. Smith songs). So I also didn't absorb the common evangelical sense of the "fault lines" that defined our culture.

However, when I became a Christian at the age of eighteen, I quickly made up for this. I drank up the Bible and consumed whatever mode of Bible teaching I could find (I'm old enough, I confess, that most of this was from huge catalogues of cassette tapes by noted Bible teachers). I abandoned my plans to become an architect, immediately sensed a call to ministry, and enrolled in Bible college. My first year at Bible college was a veritable boot camp in what I would only later learn to describe as "the culture wars."

Perhaps surprisingly, it was at Bible college that I was first taught to care about science. That might strike some as odd, since we often perceive Bible colleges as anti-intellectual zones of hostility to science. But that picture needs to be corrected a bit. In my Bible college experience, I was energized by a new interest in science bequeathed to me by the energy and passion of my apologetics teacher. A former naval engineer with a PhD in chemical engineering, "Dr. Dave" had experienced a radical conversion and also sensed a call to ministry. After spending time at seminary, he devoted himself to a teaching ministry that eventually landed him at the Bible college where his responsibilities were apologetics and "Christian evidences."

His passion and knowledge were infectious. I soaked up his fascination with archaeology (a historical science). I was awed by his presentation of geological evidences of the flood and cosmological evidence for creation. Here I was at Bible college, being invited to think about carbon dating and the Doppler effect and the geological science of sedimentation (the volcanic impact of Mount St. Helens was always a favorite case study). As someone who had skated through high school with little to no interest in science, I would never have imagined that going to Bible college would pique my interest in everything from molecules to galaxies.

Dr. Dave noted my curiosity and began to express personal interest, taking me under his wing as a kind of apprentice. Indeed, while the intellectual component fostered my curiosity about "creation science," I think it's crucial not to underestimate the personal and pastoral factors at work here as well. In significant ways, I cared about creation science because Dr. Dave had clearly demonstrated that he cared for me. I was open to being intellectually convinced precisely because I had already sensed that I was being pastorally cared for. My mind was open to creation science because Dr. Dave had expressed love and concern for my soul. I sensed a symbolic culmination of all of this when he gave me a personal copy of Ian Taylor's (rather infamous) book, *In the Minds of Men: Darwin and the New World Order*. It still sits on my shelf, no longer because I value the arguments it contains, but because I'm grateful for the love with which it was given.

Not until later did I realize that, in my Bible college education, science was primarily of interest as ammunition in a culture war. I don't mean to suggest there wasn't genuine interest or curiosity in features of God's creation and the intricacies of the physical world. I only mean that this curiosity was circumscribed and selective and instrumentalized. Science was of interest insofar as it contributed "evidences" that would help win an argument, defeat an opponent and shore up a "position" in the culture war. Science was not entertained as a vocation or calling for Christians. Instead, science was something we could *use*—and use as a weapon.

In addition, it gradually became clear to me that the "science" I was being offered was a very selective sampling of data and evidences that exhibited a kind of confirmation bias: unlike the sort of open curiosity—and openness to being *wrong*—that characterizes genuine scientific exploration of the physical world, my teachers were primarily interested in science that confirmed a certain reading of the Bible (specifically, a young-earth creationist reading of Genesis). I started to get an inkling that maybe I hadn't gotten the whole story—that maybe there was a lot more to science than flood geology and critical questions about carbon dating.

Interestingly enough, the seeds of my critical distance from this sort of "science" were also sown at the same Bible college—through an encounter with Christian theologians associated with "Old Princeton." (In Book VIII of Augustine's spiritual autobiography, *The Confessions*, he recounts his conversion through his encounter with several important books. My "conversion" with respect to faith and science is also a history of encounter with important books. Who knew libraries could be evangelists?) In some of my courses in systematic theology, my professors regularly referred to the rich heritage of Reformed thinkers that included B. B. Warfield, Charles Hodge, A. A. Hodge and others. Being a Bible and theology geek, I scoured the college library for anything and everything by these august scholars and Bible commentators. I camped out in the basement library for hours on end surrounded by their works. Whenever I could scrabble together a few dollars, I added another Warfield or Hodge to my growing personal library. My "upstairs" education in the classrooms of the Bible college were supplemented by a "downstairs," parallel education in "Old Princeton" Reformed theology. And in their work—already in the 1800s—I found quite a different posture toward science.

This all crystallized when I hit upon Mark Noll's excellent anthology, *The Princeton Theology 1812–1921: Scripture, Science, and Theological Method from Archibald Alexander to Benjamin Breckinridge Warfield*. Here I first encountered the writings of orthodox, conservative, Reformed evangelicals who were open to—and affirmative of—developments in evolutionary science. Indeed, Warfield had been cited by my professors as one of the great defenders of biblical inerrancy; but they hadn't told me about his very favorable stance towards evolution. And so some of my former sureties began to crumble. I began to sense that science was bigger than what I had been taught, and that evangelical

Christians need not be characterized by fear or a posture of defense, but could be open and curious about new developments. Most importantly, I began to realize that science need not just be an apologetic weapon. Scientific exploration could be a good in and of itself, even if that exploration might take us into places that could be unsettling.

This season in my life was a turning point in many ways. In particular, it was at this juncture that my pilgrimage in faith took me toward the Reformed tradition. (I discuss this in more detail in my little book *Letters to a Young Calvinist: An Invitation to the Reformed Tradition.*[1]) This had repercussions for every sector of my thinking, including how I thought about science. But what I absorbed from the Reformed tradition was also a stance toward history and the historical riches of the Christian tradition. The Reformation was a renewal movement in the church catholic that was birthed by the Reformers' recovery of ancient Christian sources, mining the wisdom of church fathers like Augustine and Chrysostom. That means the Reformed tradition is characterized by a sense of chronological deference, in a way—a sense that we have much to learn from what has gone before, even a certain healthy skepticism about theological novelty.

This sensibility dovetailed with my encounter with another important book in my pilgrimage: Ronald Numbers's *The Creationists: The Evolution of Scientific Creationism*. In this meticulous history, Numbers demonstrates the utter novelty of young-earth creationism as a biblical hermeneutic (a direct parallel to the utter novelty of dispensationalism as a way of understanding the eschatology of Scripture). Because my pilgrimage in the Reformed tradition had instilled in me a sense of indebtedness to the riches and legacy of the historic Christian faith, the newness and novelty of "scientific creationism" gave me serious pause. And I began to realize that the way I had been

taught to read the Bible alongside selective presentation of scientific data was, in fact, quite aberrant in the history of Christianity—a modern hermeneutical invention that was strikingly
different from the way the Bible had been read from Augustine
to John Calvin. So in a way, it was discovering the orthodox
voices of Augustine and Calvin and Warfield that made me suspicious of the notion that I needed to be a young-earth creationist in order to be orthodox.

In my pilgrimage, the riches of the Reformed tradition and
heritage have encouraged an expansive, positive affirmation of
science. This is forcefully articulated in Abraham Kuyper's 1898
Stone Lectures (delivered at Princeton Seminary), particularly his
fourth lecture, "Calvinism and Science." There Kuyper showed
how the Reformation gave rise to scientific exploration. This was
rooted, he said, in the Reformed affirmation that "every square
inch" of creation belongs to God, which means that our cultural
labor—whether as farmers or pharmacists, entrepreneurs or en-
tomologists—can be legitimate expressions of Christian vocation,
a way to serve the Lord of creation. Because the Reformed tradition encouraged scientific exploration, and because giants of
the Reformed faith like Hodge and Warfield didn't seem to balk
at affirming evolutionary science, I could say that my pilgrimage
to the Reformed tradition instilled in me an openness to evolution
I couldn't have imagined before. I reached a place where forbears
in the faith led me to see that orthodox Christian confession did
not require a narrow biblical hermeneutic on such matters. Or
perhaps I could say that my pilgrimage to the Reformed tradition
actually helped me distinguish matters of historic, essential orthodoxy from secondary matters on which Christians might disagree. The anchor and guide of the historic Christian faith pre-
vented me from inflating a theological innovation like young-earth
creationism to the status of orthodox litmus test.

However, there remained one more important phase of my own "evolution"—my own (ongoing) pilgrimage on the intersecting path of faith and science. The examples of historic figures like Augustine and Calvin and Warfield had helped me see that orthodox Christians could hold a range of positions on creation, evolution and human origins. And so the tent of the faithful was enlarged beyond the small circle of young-earth creationists. It was less a matter of having changed my position and more a matter of recognizing that a range of positions could be consistent with orthodox Christian confession.

However, I noticed that not all of my colleagues shared this "big tent" sensibility. Instead of being characterized by openness to a range of positions, some convinced proponents of evolutionary creation looked rather familiar to me. Indeed, I began to realize that while some of my friends and colleagues who were evolutionary creationists affirmed a very different sort of science from what I'd been taught at Bible college, they actually *mirrored* my Bible college professors insofar as they were *using* science in a similar way. While they had swapped positions (and many of them had been former young-earth creationists), they hadn't given up the culture wars stance that comes with such positions. Science was still a weapon used in a war. The point was *winning*, not witness.

And it seemed to me that this stance was fostered by fear. If young-earth creationists feared the erosion of biblical faith and a compromise of the gospel—a fear that drove their "culture war" stance—then some of my evolutionary creationist colleagues seemed to fear being perceived as hicks and fundamentalists, losing the respect of their colleagues in the academy or opinion-shapers in culture. In each case, it seemed that these very different fears nonetheless occasioned similar responses: brandishing "science" as a weapon in order to *win*.

While it might have been (some) Christians who taught me to fight, the Spirit of Christ has taught me to "be not afraid." This means I don't need to fall into fight-or-flight mode, driven by fear to either crush all threats or hide from difficult questions. It also means that I don't need to confuse a particular position with the *only* way to follow Jesus. While Christians need to wrestle with and work through difficult questions at the intersection of faith and science, that intellectual labor needs to be rooted in the core conviction that all things hold together in Christ (Colossians 1:17).

As my friend and colleague Michael Gulker of the Colossian Forum sometimes puts it: true discipleship is learning to "fight like Jesus." That, of course, is an ironic, provocative way to remind us that if we look at the model and exemplar of Jesus, we'll see that he wins by losing. As N. T. Wright provocatively puts it in *The Challenge of Jesus*, "The cross is the surest, truest and deepest window on the very heart and character of the living and loving God."[2] So in all of our cultural labor—including science, theology and the conversation between the two—"the methods, as well as the message, must be cross-shaped through and though."[3] By bearing witness to the coming kingdom, we—like Christ—will be witnesses by giving up our penchant for winning, for domination and control—either of the culture or of our brothers and sisters in Christ. To follow Jesus is to beat our swords into plowshares and our microscopes into vessels of worship and praise. It is also to prize our common witness above being right or respected. This means refusing to instrumentalize science for either party in the culture wars by eschewing such warring altogether, following instead the Prince of Peace, in whom all things hold together.

- 2 -

Who's Afraid of Science?

Scot McKnight

Scot McKnight *is the Julius R. Mantey Professor of New Testament at Northern Seminary. He completed his PhD at the University of Nottingham with a dissertation on the Gospel of Matthew. A native Illinoisan, Scot and his wife, Kris, live in the northern suburbs of Chicago, and they have two adult children and two grandchildren. They enjoy gardening, walking and travel. He is the author of more than fifty books.*

■ ▪ ■ ▪ ■

A STUDENT OF MINE RECENTLY WROTE A PAPER in which he talked about growing up in a church that taught young-earth creationism. This student had begun to feel quite uncomfortable, as he had learned some facts about the world a long, long time ago that were almost certainly undeniable. Here is how he reported it to me:

> Specifically, I remember the single youth group lesson that finally pushed me to the point of crisis. We were being taught about creation from a perspective that I now know to be called young-earth creationism. I remember watching the

video our youth leader and pastor had selected, and as they launched into the prefabricated curriculum, I remember raising my hand and asking a single, simple question: "What about the dinosaur bones?" The pastor and youth leader looked at each other, exchanging some unspoken communication, and then our pastor looked me in the eyes and said "Satan buried those bones." After receiving comparably absurd responses to questions I asked about carbon dating and human archaeological evidence, I walked away feeling deeply shaken and concerned.

When we are driven to think that dinosaur bones were buried by Satan to fool the world into demonic ideas, and to think the whole world is duped, and that we alone are right in our interpretation of the Bible, I contend that we need more humility, not more confidence. We need enough humility to give the Bible yet one more read to see if we've got it right.

I grew up in the environment of young-earth creationism, though because my childhood pastor had been to a university and not to an encapsulating Christian college, we were less strident about our beliefs. Yes, we firmly believed God created the world, and yes, we believed that evolution was rooted in unbelief and often enough in atheism. But we did not develop the kind of explanation my student heard. My own Christian college experience, while never once given to such explanations, did mean a robust, evidence-based defense of creationism. As a college student I thought I had expanded my brain to the breaking point when I read Francis Schaeffer's *Genesis in Space and Time* and began to think the cosmos just might be older than I had been taught.

Later, as a seminary student, I read L. Duane Thurman's *How to Think About Evolution and Other Bible-Science Controversies,* a much more nuanced, science-respecting but theologically alert volume. After that I became more and more accustomed to one very simple dimension of thinking: *Base what I believe on the evidence.* Oddly enough, l learned this from my fundamentalist Bible college teachers, who taught me over and over—yea, they preached it—to read the Bible for myself, to find the evidence, to sort out the evidence, and to base my theology on the evidence and the evidence alone.

This is the hermeneutical equivalent of the *scientific method.* Not once was I taught to believe something because the church taught it, to believe something because the creed said so, or to believe something because it was the inherited tradition. To be sure, there was some inevitable tension if I were to disagree with the tradition I was in. For instance, I became obsessed with the rapture question in college, read a bundle of books—more than I care to confess—and then came upon George Ladd's famous defense of what we called the "post trib view" (George Eldon Ladd, *The Blessed Hope*) and became his advocate. My Bible college professor actually told me I wouldn't be permitted in his class on eschatology because—this is what he told me explicitly— (1) I knew too much about the issue and would have too many questions others would not have, and (2) I was too much in agreement with Ladd.

I took this as a badge of merit, because I knew I was doing exactly what the professor had taught me—*basing my beliefs on the evidence of the Bible and the Bible alone*—and because I would get credit for taking a class by reading and not showing up for class. (I was tempted to barge in but he told me to stay away. We met a number of times—I got to air my views and he humored me by pushing me harder.)

What happens when we apply this approach *both* to the Bible (as I had learned) *and* to the question of origins? We learn to base what we believe—about the Bible, about origins, about age—on the evidence and the evidence alone. Over the next decade of my life, I came to believe that if I was going to base my faith on the evidence of the Bible, by examining it and challenging as well as affirming the church's beliefs, then I had to be honest and fair to do the same for questions about the age of the universe, the age of the earth and the question of origins.

The result was conflict between what I had been taught about origins and age—even if I allowed the earth to be as old as twenty thousand years, which I thought was mighty liberal of me—and what I was reading in the Bible (as I had been taught to read it). The word "conflict" is probably too mild. At times I came to the conclusion that *my Bible* might be wrong. Then along came a series of very encouraging books and articles and conversations, too many to mention in this context, that provided another way. I learned that "my Bible" was in fact *my reading of* the Bible. Maybe it wasn't so much the Bible that was wrong but *the way I was reading the Bible* through the lens of my own questions—questions shaped more by my past worries about evolution and less by learning to read the Bible in its own historical and theological context.

I have never read all of Charles Darwin's *The Origin of the Species*, though I have dipped into it, read tons about it, and learned both its strengths and weaknesses. While reading Adrian Desmond and James Moore's hefty *Darwin: The Life of a Tormented Evolutionist,* I settled rather gently into comfort with the general orientation of evolution. Since those days, nearly twenty years ago now, I have read book after book that has helped me think more critically about science and evolution, none more helpful or accessible than Edward J. Larson's *Evolution: The Remarkable History of a Scientific Theory.*

My own field is not science but the Bible. Learning about science has taught me to be more scientific about the Bible, not less. It has taught me not to succumb to simplistic theories about the Bible, not to settle for less than rigor about what Genesis 1–3 are saying, and not to force an ancient Near Eastern text (Genesis) into the thought patterns and categories of modern science. Learning about science has taught me humility in my Bible reading and has pushed me to think again, to read again, to ask again, and to wonder again what the Bible was saying when it was written and how the Bible was heard by its original hearers (so far as the evidence permits us to know such things).

What science has taught me, then, is that there is no reason to fear science. It has no agenda. Science, in its best form of study—whether it's examining the Bible or the universe—does not impose; it looks. It asks the evidence to talk to us and it lets the evidence make the decisions. It asks the observer—again, the Bible reader or the universe examiner—to get out of the way to hear, to watch and to record what is there. It asks us not to be afraid but to respect what is there.

Science, then, encouraged me to think again about the Bible. One recent study, by John Walton, has urged us to think again about what Genesis 1 meant in its day.[1] Some elements of Walton's theory are being challenged, but his major idea is reasonable and persuasive: that Genesis 1 is not about the origins of the world so much as the function of God's world. That is, Genesis 1 presents God fashioning the world as his temple, placing us in it to reflect his glory and to govern his good world on his behalf. That is, the universe is God's temple and we are summoned by God to care for God's temple by worship and work.

The scientific method encouraged me to look at the Bible more freshly and gave me the courage to listen to science enough to rethink what the Bible might be saying. Science, then, has given me renewed confidence in our ability to hear what God is saying in the Bible.

3

The Inevitable Conclusion

Ken Fong

Ken Fong *has been a pastor since 1981 of Evergreen Baptist Church of LA in Rosemead, California, which he has led to become a multigenerational, multiethnic, majority Asian Pacific Islander-American congregation. He is currently the executive director of the Asian American Initiative at Fuller Seminary (Pasadena) and assistant professor of Asian American church studies. He resides in Sierra Madre with his wife, daughter and two dogs.*

■ · ■ · ■

I DON'T KNOW WHAT YOU USE SWIMMING POOLS FOR, but I don't remember doing much swimming in them when I was growing up. Instead, we invented games that would work in a watery playground. One of the more challenging ones was to clutch a beach ball in one arm and then dive down to the deep end with it, trying to place the buoyant ball on the drain and hold it there. The deeper we went, the more the ball would want to escape our grasp. The ball would never stay buried for long. It belonged on the surface. Despite our best efforts at holding it under, its emergence back on the surface was inevitable.

I grew up in a Chinese-American home where education was stressed alongside going to church and studying the Bible. Like most kids (back then), I developed a fascination with dinosaurs and prehistoric mammals. Even as I was memorizing the Latin names for all my favorites, I was also memorizing Bible verses and learning to understand everything through a literal reading of this holy book. In ninth grade, in preparation for my baptism, I met with our pastor.

"Ken, do you have any lingering questions that you'd like to ask before you make this lifetime commitment to Christ?"

I replied, "Well, I really only have one question for you. If you can clear up my confusion, I think I'll be fine getting baptized. Okay, so how do you explain the dinosaurs and the six-day creation story in Genesis?"

After the briefest of pauses, my pastor simply said, "The Bible tells us that with God, one day is like a thousand years."

"Okay then. I'm ready to get dunked."

In fact, though, his response really didn't extinguish the glowing embers of my doubts about this matter. But I *acted* like it did. Not just there in his office, but in all kinds of different settings over the next two decades. I was a biological science major (pre-med/dental) at UC Berkeley in the early seventies, where I inwardly struggled with the growing contradictions between what scientific evidence pointed to and the way I'd been taught to interpret the Bible. Looking back, it's clear to me now that my primary issue wasn't the veracity of the science so much as the veracity of the Bible. Whenever I sensed that the Bible's authority was at risk, I would unleash a barrage of attacks against the threat to discredit and dismiss it, and I would submerge the disturbing thoughts and feelings in the deep end of my mind. I devoured every book I could find in Christian bookstores that promised to substantiate my need to believe that God created everything and everyone in six twenty-four-hour days. During college and over the following few years, I equipped the students in our church and at denominational camps with those same talking points, underscoring the scientific accuracy of the biblical account of creation. In retrospect, I think I was a closeted evolutionary creationist who was scared to death of coming out.

After I earned my bachelors degree, I gave in to the crazy notion that the God of the universe was calling me to be a pastor. So off I went to seminary, still using way too much energy to suppress my inner conflict with young-earth creationism. In my Old Testament survey class, I purposely chose to argue for the scientific basis for the Noachian deluge: There must have been a time in the earth's distant past when giant reptiles, sizable mammals and early humans all thrived post-Garden in a nearly worldwide tropical rain forest. But due to the visitation of a sizable planetoid, the resulting seeding of the atmosphere and severe tidal conditions triggered an unprecedented global deluge. Most of the dinosaurs weren't taken aboard the ark and thus drowned, as did everything else except for Noah and his family and the paired specimens he saved. As the floodwaters finally receded into the ocean basins, countless numbers of drowned creatures were carried along by retreating rivers, with those of similar mass being deposited in huge caches along the way. Oh, and of course, due to the magnetic poles, many wooly mammoths living near the North Pole were flash frozen, some found today with tiny, intact buttercups in their mouths. I'm pretty sure I included a section that debunked the systems used in the late seventies to arrive at dates of fossils, too.

Somewhere doing my time in seminary, I also latched on to the intelligent design (ID) approach. This way of looking at things, which started with "logic" instead of trying to defend the veracity of the biblical account, seemed to be the missing piece in my presentation. But I soon ran headfirst into an ID wall. If this approach was supposed to be about scientific evidence that should logically cause even the most skeptical person to believe that everything must have been intentionally designed by one super-smart deity, then why did it start to sound more like apologetics than science? One of the basic tenets we science majors were taught was that any and every theory we devise should be falsifiable. In other words, genuine scientists don't just devise experiments that are guaranteed to substantiate their hypotheses. They must be open to coming to conclusions, based on their experiments, that seem to contradict what they think is true. Nowhere in my presentation of the ID approach was there even the slimmest chance that some of the evidence would point *away* from the belief that God exists, and that God brought everything into existence. No wonder few science-trained non-believers were ever motivated to come to faith in Christ through my enthusiastic diatribes on ID.

A pastor I served with had also been a life science major. One day when no one else was around I asked him if he believed in the theory of evolution. "Sure, well, at least within species. But I would never say that from the pulpit. When I'm preaching, I will always promote young-earth creationism and intelligent design."

I stammered on. "But, if you actually see evidence for God using some forms of evolution, why would you never talk about that?"

Confidently, he replied, "Because I don't want to confuse people."

As I walked slowly out of his office, I muttered under my breath, "Yeah, well, *not* mentioning that you believe some parts of Darwin's theory is going to confuse a whole bunch of the science-trained members who simply can't reconcile a literal reading of Genesis with what science has discovered so far."

Surprisingly, that turned out to be a major turning point for me on this issue. As I mulled over my colleague's clear decision to give everyone the inaccurate impression that he was a creationist so folks wouldn't be confused, I ultimately decided that I was going to surface my doubts and embrace solid science, even if it cost me. Twenty years later I'm the senior pastor and I've done my best to fulfill my vow. I began by reframing the Garden account in Genesis whenever a sermon would land there: "Most of us grew up reading this as a blow-by-blow account of how God created everything. But as I've sat with this text all these years, I've come to see that this passage is really about the *who* and the *why* rather than the *how* and the *when.*" I'm sure that one of the evolutionary creationists I used to argue with back in seminary told me that, which didn't cause me to see things his way. Nevertheless, I regularly lay out that same line of thinking about Genesis 1 and 2. Not because I think I'm able to say it more convincingly, but because it's now what I believe. When I think back to my ongoing struggle over the years to integrate my Christian faith with my scientific mind, I'm no longer surprised at what has finally come to the surface of my thinking.

One of our deacons has also had a huge impact on my thinking and my boldness. When he went to UCLA to study astronomy, he wasn't a Christian. But through the evangelistic efforts of a large, conservative church, he gave his life to Christ and began attending that church for the next seventeen years, even after he'd earned his PhD at Cal Tech and went to work for the Jet Propulsion Laboratory. When he showed up not long ago at our church, I asked him why he left that church, which had birthed his faith. "Don't get me wrong. My logical brain loved all the emphasis on correct doctrine and systematic theology. But whenever the sermons would touch on the origins of life and the age of the earth, I felt like jumping to my feet and screaming, 'You don't know what you're talking about! The evidence proves that the earth is about 4.5 *billion* years old, not 6,500.' But no one seemed interested in asking a JPL scientist who was a fellow believer why he believed it wasn't scientific to date the earth by counting the generations listed in the Bible. We loved each other, but I needed to find a church that respects the truth that God has revealed outside of the Bible, too."

So when it came time to preach a message about the creation of the universe, it was obvious that I needed to bring up this brilliant scientist and deacon to share the pulpit with me. With high-resolution deep space photos taken by the Hubble Space Telescope on the wide screen above our heads, we recited Psalm 19:1, "The heavens declare the glory of God." Then I had him share why, as a world-renowned planetary scientist, he saw no contradiction between believing that Jehovah God created the universe and the evidence that our universe was formed through the mind-blowing process we have come to call the Big Bang. When that message ended, we projected the lyrics of our closing worship songs over other photos taken by Hubble. Just hearing the enthusiastic reaction from a variety of our science professionals afterwards fueled my ongoing conviction that I can help bring together two critical parts of our people's thinking: their faith in God, and their faith in the ongoing work of faithful scientists who are committed to uncovering truth that may or may not be found explicitly in the Bible, but is God's truth nonetheless. Some did leave our church, because it made them extremely uncomfortable to hear that, although our faith inspires us to believe that God made everything, the mountains of evidence indicate that God used the Big Bang and evolution to accomplish this miracle. But I am convinced that many others will hold onto their faith or even come to faith upon hearing that there's definitely a place for robust science in the search for God and God's truth.

4

Learning to Praise God for His Work in Evolution

Deborah Haarsma

Deborah Haarsma *is president of BioLogos. She received her PhD in astrophysics from the Massachusetts Institute of Technology. She and her husband, Loren, wrote* Origins: Christian Perspectives on Creation, Evolution, and Intelligent Design *(Faith Alive Christian Resources, 2011). While she and Loren both enjoy science fiction and classical music, only Deb enjoys gardening.*

■ · ■ · ■

I WAS SITTING AT THE KITCHEN TABLE with my dad, my high school biology textbook lying open in front of us. The public school biology class had begun the unit on evolution, and I had a lot of questions. My parents had always encouraged my interests in science and math, from enrichment classes after school to math games at home. They also modeled the Christian faith for my brother and me and as a family we were active in a devout evangelical church. I learned to love Jesus and the Bible from an early age and committed my life to Christ, and as I grew to adulthood that commitment was renewed several times.

Everyone at church and at home thought that the earth was young and evolution never happened. This view wasn't presented in a dogmatic way, but what other Christian view could there be? At school, I was fortunate that my biology teacher didn't teach evolution as some sort of antireligious propaganda, but he couldn't address the biblical concerns I had either. So there I sat with Dad, telling him about what was being taught in class about evolution. The evidence and arguments in the textbook were a lot more compelling than I expected, but surely evolution was atheistic and contrary to the Bible? We tossed the ideas back and forth for a while, since this was new to both of us. Finally my dad sat back and said, "I don't know." I still remember the sense of *relief* I felt. It was okay to not know! If adults didn't have it all worked out, then surely as a teenager I didn't have to come to a conclusion right away. It was clear that my dad's belief in God wasn't threatened by the conversation, so I was assured that my faith wouldn't stand or fall based on what I decided about evolution.

During college I put off the questions about origins in my preoccupation with choosing a major. Which career would serve God best? I attended a Christian college, Bethel University in St. Paul, Minnesota, and it was there that I fell in love with physics. I know physics isn't for everyone, but I loved it. I remember working with a group of students in the freshmen physics lab, trying to put an experiment together and figure out what to measure in this messy, real world. We did the mathematical calculations that were supposed to describe the experiment and then came the moment of truth: we compared these predictions to the real world data . . . and it matched!

I was hooked. I had experienced firsthand how math actually describes the natural world. Physicist Eugene Wigner, a Nobel Prize winner, called this the "unreasonable effectiveness of mathematics."[1] In physics class that day, I encountered a God who crafted a universe of such order and regularity that it can be described with reason and logic. And I experienced an aspect of being created in the image of God. God gives us the ability to understand something of how he governs the world. I saw how science could be a Christian vocation, how I could be a Christian with all of my mind as well as all my heart.

After graduation I moved to Boston to pursue a PhD at the Massachusetts Institute of Technology. After the close-knit Christian learning community at Bethel, this major international research university was a bit of a shock. But it was also an amazing opportunity to work with some of the best scientists in the world. I became interested in astrophysics and discovered that physical laws like gravity and magnetism, which we understand pretty well on earth, manifest themselves in dramatic extremes out there in the universe, such as black holes and neutron stars. Astronomy showed me that the Creator has filled the universe with an abundance of wonders, wonders that I was eager to share with other believers so that they, too, could join me in praising God.

Once I started studying astronomy, however, I quickly encountered the scientific evidence that the earth and the universe were more than a few thousand years old. I couldn't ignore the origins question any longer. I took a hard look at the scientific evidence, but the more I looked, the more solid the evidence proved to be. There wasn't just *one* way to measure age, but several independent measurements and arguments that all pointed to the same general conclusion: billions of years, not thousands.

This drove me back to the Bible. I had to wrestle with how to understand the Bible that I loved. What was Genesis really teaching? This was many years before BioLogos began, so answers were harder to track down. But I discovered books about the age of the earth and Genesis, where I could read Christian astronomers and geologists describing the scientific evidence and not be afraid that they would put an atheistic spin on it. And I found a wonderful community of fellow Christian graduate students in my InterVarsity chapter who were discussing the same issues. I knew I wasn't alone.

As I began to read the work of biblical scholars, I learned that the ancient Egyptians, Babylonians and Hebrews believed the earth was flat, with a solid dome sky and an ocean above the sky. That picture sounds totally strange to us, but they really believed it! They thought that rain falls when holes open up in the dome to let the water through. But this ancient picture helped me finally understand what's happening on day two of Genesis 1. God said, "Let there be a firmament to separate the waters above from the waters below." It dawned on me that God didn't try to correct their misunderstanding of the scientific picture. He didn't try to explain atmosphere and evaporation and precipitation. Instead, God accommodated his message to his people's limited understanding so that they could focus on the main points: the world is not filled with many gods, but is ruled by one sovereign God. Creation is good and humans are very good, bearing God's image. I came to believe that these are the primary messages of Genesis to us today—that the Bible is much more concerned with the *who* and the *why* of creation, while the universe is where God left clues about the *how* and the *when*. I came to accept that the earth was old and that God created the universe billions of years ago in the Big Bang.

It took more time for me to accept evolution as the means by which God created the diversity of life. I was reluctant to shift my views even farther from the views of my childhood Christian mentors. But after reading what Christian biologists had to say, I realized that the evidence for evolution was as strong as the evidence for an old earth. Like the evidence for age, the evidence for evolution comes from many types of measurements all pointing to the same general picture. It's not just about fossils, but about anatomy and the distribution of creatures around the globe. Moreover, I saw a dramatic pattern of prediction and confirmation—Charles Darwin introduced evolution before anything was known about DNA. The predictions of the evolution model have been dramatically confirmed in recent decades by discoveries in genetics. DNA evidence provides abundant confirmation that all life forms on earth are related by evolution through a tree of common ancestry. I came to see evolution as a scientific description of how God created the species.

However, intellectual acceptance of the evidence wasn't the whole story. This picture of the world shook me up. It was very different from how I was used to thinking about God's creation, and it took another few years for me to repattern my worship habits to match it. For instance, what should I think about while singing hymns like this one?

All things bright and beautiful, all creatures great and small,
All things wise and wonderful—the Lord God made them all.

Each little flower that opens, each little bird that sings—
He made their glowing colors, he made their tiny wings.

The purple-headed mountain, the river running by,
The sunset, and the morning that brightens up the sky.

All things bright and beautiful, all creatures great and small,
All things wise and wonderful—the Lord God made them all.

(Cecil F. Alexander, 1848)

During my whole upbringing, I sang hymns such as this while picturing God walking through the Garden of Eden with each bird flying out of his hand in a separate, special miracle, or picturing C. S. Lewis's Aslan singing the rocks and hills into being. But now I had come to believe that God made these things through natural processes over millions of years. If God made birds and mountains using evolutionary biology and tectonic plate motion, what do we praise God *for*?

Over the years, I have found many good answers to that question, some of which I explored in a book I wrote with my husband, Loren Haarsma, called *Origins: Christian Perspectives on Creation, Evolution, and Intelligent Design*. I now praise God for *working over the long term*. For example, when I sing hymns about God creating mountains, I picture God using the convection in Earth's mantle to slam the Indian sub-continental plate into the Asian continental plate, a very slow but very mighty push to raise up the snowy heights of the Himalayas. The time scales of this process are staggering, and I'm reminded to praise the God for whom "a day is like a thousand years" (2 Peter 3:8). God's time is far beyond ours.

I also praise God for *the glory of the system*, in addition to each individual thing in the system. Not only did God make each individual mountain, but God carefully designed a whole system to make all the mountains on earth. When I sing hymns about God creating flowers, I think of the evolutionary mechanisms God designed. Evolution produces not just a few kinds of flowers, but an extravagant abundance of flowers with every variation of size, shape, color and scent. God designed a system that creates abundant beauty. Even the language of Genesis 1 resonates with this. On day three, God says, "Let the land produce vegetation," rather than "Let there be vegetation" (Genesis 1:14)—God is working through the system of the land to bringing about every seed-bearing plant.

And I now praise God for *his faithful upholding of the natural world*. The continual functioning of the natural world, day after day, year after year, and over billions of years, is a tremendous testimony to God's faithfulness. In fact, in Jeremiah 33:25, God points to the regularity of day and night and the "established laws of heaven and earth" as evidence of how faithful he will be in keeping his promises. Without God's sustaining hand, the laws of physics would cease to work, matter would disintegrate, energy would disappear and the very fabric of space and time would dissolve.

I've learned to praise God for *what is glorious in his creation*, even if it isn't miraculous. God is just as present in the regular workings of nature as he is in supernatural acts. It's tempting to say, "The universe is so amazing, even scientists don't understand it!" This implies that God is best seen in what science can't explain. But a scientific explanation does not replace God. Science gives us a human description of *how* God is creating and sustaining. Far from wiping out a spiritual view of the universe, a scientific explanation can actually increase my wonder and awe, as I get a glimmer of how God makes it work.

Finally, I've discovered *aspects of creation that illuminate Scripture*. We live in a stunningly immense universe—our own galaxy contains billions upon billions of stars, and it is just one of billions of galaxies in the universe. That can make us feel very small. Astronomer Carl Sagan gave an atheistic perspective on this, writing in his book *Cosmos*, "We find that we live on an insignificant planet of a humdrum star lost between two spiral arms in the outskirts of a galaxy which is a member of a sparse cluster of galaxies, tucked away in some forgotten corner of a universe in which there are far more galaxies than people."[2]

Insignificant? Forgotten? The Bible looks at the same universe and tells a very different story. Psalm 103 tells of the vastness of the creation with phrases like, "as high as the heavens are above the earth." But the Psalm doesn't go on to say, "Humans are tiny and insignificant before God." Rather it says, "For as high as the heavens are above the earth, so great is his love for those who fear him; as far as the east is from the west, so far has he removed our transgressions from us" (vv. 11-12). When we look at the vastness of universe, God doesn't mean for us to feel insignificant. He means for us to see the vastness of his love and forgiveness.

I haven't worked out the answers to all of my questions about evolution. On some points I still say, "I don't know." But I've learned to praise God for it. In evolution, God worked faithfully, in his own time, designing systems to create a glorious world filled with abundant life. A world that, when seen through the eyes of faith, reminds us of the vastness of his love for us.

5

An Old Testament Professor Celebrates Creation

Tremper Longman III

Tremper Longman III is the Robert H. Gundry Professor of Biblical Studies at Westmont College. He has a PhD from Yale University in ancient Near Eastern languages and literature. He is married to Alice and has three adult sons and two granddaughters. For fun and exercise, he plays squash.

■ · ■ · ■

REFLECTING ON THE DEVELOPMENT of my thinking about the question of the Bible and the science of origins takes me back to the time I first committed my life to Christ. When I was in high school I wasn't a Christian or, to be honest, much of a student. Though I became a Christian the summer before I started college, an interest in academics didn't come with my newfound faith. During my freshman year in college at Ohio Wesleyan University, however, I found my faith increasingly challenged, particularly by the Religious Studies department. I also met a girl who had recently become a Christian through the ministry of students from Westminster Theological Seminary, near her high school in Philadelphia. These students were not bound for ministry but for promising academic careers, so they devoted not only their hearts but also their minds to the service of the gospel. They encouraged her to think seriously about her Christian faith. I met her right after she spent a summer studying with Francis Schaeffer and a young Os Guinness at L'Abri in Switzerland, and her intellectual interests were contagious.

In short, the challenges of my professors and the personal challenge of my future wife, Alice, drove me to take my academic life more seriously. As I did, my intellectual curiosity about the faith grew and grew. During college at Ohio Wesleyan (1970–1974), I majored in Religious Studies and took many courses in philosophy and the other humanities, but only one class in the sciences (in astronomy). I have little memory of thinking seriously about the issues of faith and science in college, though I do remember a local minister debating some of our science teachers. In retrospect, the minister, who had no business debating academics, must have been a concordist, since he argued that "Job invented electricity." If anything, this experience discouraged me from thinking further about science and the Bible. It was just too embarrassing.

After college I went to seminary at Westminster in Philadelphia (1974–1977) where Alice's spiritual mentors had studied. At that time the Old Testament professors didn't devote a lot of time to Genesis 1 and 2, but they did argue for a more figurative understanding of these chapters than many other conservative scholars (including the "framework hypothesis"). I don't remember them attacking evolutionary theory as evil or even wrong, though at that time they defended the idea of a special creation of an original historical couple, Adam and Eve.

After seminary I headed to Yale University, where I got my degree in ancient Near Eastern languages and literature (1977–1983). It was here that I was exposed firsthand to the great creation myths of Sumer, Babylon, Egypt and Canaan, noticing right away their dramatic similarities and clear differences with Genesis 1–2, as well as the many other creation accounts of the Old Testament (Psalms 74, 104; Job 38; Proverbs 8 and so on). This exposure would prove to have important ramifications for the deepening of my understanding of the Bible's creation account.

After Yale, Westminster hired me, so I returned to the school where I had done my seminary degree. During my tenure in Philadelphia (1981–1998) there were many controversies, but none really on creation. Everyone pretty much took the view that there was considerable figurative language in the early chapters of Genesis (especially the "days") and they taught that the earth and the cosmos were old. Even so, as was true during my student years, most if not all of the faculty, myself included, affirmed the special creation of Adam and Eve. At the time, it seemed critical to an Augustinian interpretation of Romans 5, which linked our sin nature to Adam's sin in a way that suggested a hereditary connection.

The next phase of my journey in considering Genesis and cosmic and human origins came when I wrote a book on Genesis (*How to Read Genesis*). I didn't engage the question head on in the book or devote much space to it, but throughout I advocated the importance of reading Genesis as a whole, including the creation account, in the light of its ancient setting: "The important point that comes to the fore through this kind of study is that the Bible is a literature of antiquity and not modernity. This truth will have a great impact on our study. For instance, we will come to realize that the biblical creation accounts were not written in order to counter Darwinism but rather the *Enuma Elish* and other ancient ideas concerning who created creation."[1] Already by the time of the writing of this book, I had come to understand that Genesis 1–2 were interested in celebrating the fact *that* God created the heavens and the earth and all that live in them, including humankind—but not in *how* he did it.

The event that really propelled me into the present controversy over the question of origins came out of nowhere. I was
teaching a small group of interested and highly educated young
adults at a retreat center on Lake Tahoe in September 2009. My
subject was "the story of the Bible"—how the parts of the Bible
all fit together to tell the drama of our redemption. I was approached by one of the participants, a professional documentary
filmmaker, who asked if he could interview me on camera for
about an hour. To be honest, I was rather exhausted from the
day's teaching and wanted to relax, but I agreed. The time turned
out to be enjoyable. He asked me a wide range of questions
about the Old Testament, including one about the historical
Adam. He asked, "If it turns out that there was no literal historical Adam and Eve, does that mean that the biblical creation
account is not true?" The question was motivated by my teaching
on the highly figurative nature of Genesis 1–2, its interaction
with ancient Near Eastern creation stories and the fact that the
two creation accounts (Genesis 1:1–2:4a; 2:4b-25) do not share
the same sequence of events, indicating that we are not getting
a literal account of God's creation of humanity. While not committing myself to the view that Adam and Eve were not literal, I
suggested that if it turns out that they were not it did not undermine the message that the biblical author intended to communicate (*that* God created humans, not *how* he did so).

I had no idea what he was going to do with this film, but soon
found out when I got an email from the dean of Reformed Theological Seminary, who had just viewed it on YouTube. I was
scheduled to teach at Reformed's Washington, DC, campus in a
matter of days. His concern was not only with the issue of the
historical Adam. He began with my view that the biblical account did not require a rejection of evolution. Within a couple
of days of that first email I was fired from my job as a regular

adjunct at Reformed. I didn't know that their board prohibits anyone teaching (apparently part-time as well as full-time) who did not believe that Genesis was incompatible with the theory of evolution. Soon after I was fired, my good friend and former colleague Bruce Waltke resigned under pressure from his more full-time position with Reformed for the same reason.

I mention this episode out of no animosity toward Reformed, a seminary I still respect a great deal, even though I believe this particular policy is shortsighted and deeply problematic. I include it because it was instrumental in getting me to think more intentionally about the subject.

Since that time, I have done much more writing and teaching on this question. I have tried to educate myself a bit on the science, though I fully recognize I am not a scientist. Fortunately, a number of helpful resources are available to help a layperson like myself, provided by Christian biologists like Jeff Schloss at my home institution of Westmont College in Santa Barbara (with whom I recently did a Veritas Forum on the subject), Dennis Venema, Francis Collins, Karl Giberson and many others. I have also benefitted from interaction at symposia and conferences with Old Testament colleagues like John Walton, John Collins, Peter Enns, Todd Beal and Richard Averbeck. We don't always agree (indeed on some points we have vehement differences), but the discussions have taken place in a spirit of collegiality, respect and common commitment to the Word of God.

Prominent among my colleagues in science is Richard F. Carlson, a research professor of physics at the University of Redlands. After reading my *How to Read Genesis*, Dick approached me about writing a book about the Bible and science and I readily agreed. We actually began work on this project before the YouTube incident, and so the project became another major contributor to my renewed interest in the subject. Our book,

Science, Creation and the Bible: Reconciling Rival Theories of Origins, came out in 2010.

I look forward to continuing to think about this important issue. Many interesting biblical and theological issues deserve renewed scrutiny. I am not a scientist, nor am I an apologist for evolution or for the idea that humanity goes back to a breeding population of some thousands of individuals, not a single pair. As a biblical scholar, though, it is important to maintain that the Bible does not proscribe these ideas. We should not criticize scientists who come to these conclusions and we should not tell our children that what they are learning in their biology and physics classes contradicts the Bible. While the biblical text does not speak to the issue of *how* God created, it does insist *that* he was the Creator. This truth is not discernible by scientific inquiry but by the eyes of faith and belief in his Word. While humanity may not go back to a representative couple, it seems to me that the biblical text does present the idea of what I heard theologian Jamie Smith of Calvin College call the "episodic nature of the Fall." That is, when human beings were endowed with God's image, they were morally innocent. Our sinful nature is not due to the way God made us, but the result of our own human rebellion.

These issues deserve careful examination. I look forward to continuing to think about and discuss these biblical and theological topics with scientists, theologians, biblical scholars, ministers and others. May we do so with prayer, devotion to God and his Word, and with respect for each other, even as we find ourselves in disagreement.

6

Embracing the Lord of Life

Jeff Hardin

Jeff Hardin *(PhD, University of California-Berkeley; MDiv, International School of Theology) is chair of the Department of Zoology at the University of Wisconsin and chairman of the board at BioLogos. His research focuses on embryonic development, using the roundworm C.* elegans *as a model. He advises several Christian student groups at UW and is an elder at his church. He and his wife have two adult sons.*

■ · ■ · ■

In February 2009 I found myself chafing uncomfortably in a coat and tie in front of TV cameras. As a lifelong academic I am no media darling, but I was unable to say no to one of my students. Besides being professor and chair of the Department of Zoology at the University of Wisconsin-Madison, I also serve as faculty director for an honors biology program called the Biology Core Curriculum, or Biocore. A Biocore student and pastor's son involved in the local InterVarsity Christian Fellowship was an intern at the local public television station. He knew I was a Christian and asked if I would be willing to be interviewed for the bicentennial of Charles

Darwin's birth and the sesquicentennial of the publication of his *On the Origin of Species.*

There were other reasons to chafe besides hot lights, of course. Some of my non-Christian colleagues at the university and wonderful Christian friends are in agreement: evolution and Christian faith are implacable foes.

Unlike many writing in this volume, I never experienced deep conflict between science and Christian faith. For me, Jesus has always been the Lord of life—Lord of eternal life (John 10:10) and Lord of the resplendent biological life teeming on this planet (Colossians 1:16-17; Hebrews 1:3). My parents had abandoned any sort of traditional Christianity when I was in grade school, and yet there were hints of God's splendor then: a book in kindergarten about Sirius (the Dog Star) that explained God was there, and family gatherings to watch nature documentaries on live TV. To me the natural world was an exciting opportunity for discovery. As a child I committed to becoming a scientist.

As a petulant middle-schooler, though, I had not given much thought to whether God existed. And then—at a Southern Baptist youth revival in 1972—I had an overwhelming sense of my need for a Savior, and in dramatic fashion I turned to Christ, coming to know the Lord as "the Giver of Life." I was baptized at the local United Methodist Church, and grew in my faith. Later, a local Southern Baptist church helped me discover the Bible. Curiously, in neither place was I taught that evolution and Christian faith were incompatible. Perhaps I didn't pick up on the implicit controversy, or perhaps the emphasis on warm, personal faith drowned out such issues.

I went off to college at Michigan State University. I wanted to be a physicist, but my struggles with solving electrostatics problems in spherical coordinates quickly disabused me of such aspirations! During the spring of my freshman year my younger brother had nearly died and was left painfully disfigured from bacterial meningitis, so I settled on a premedical curriculum and changed my major to zoology.

As is true of many students, my spiritual life had plummeted by the end of my freshman year. Thankfully, through the help of loving people in the local Campus Crusade for Christ student ministry, I began to follow Christ in earnest midway through my sophomore year. C. S. Lewis's academic bent was to my liking, and so I devoured his books. *The Problem of Pain* shaped me more than I knew. Lewis was not particularly concerned with precisely how human beings had come into relationship with God, but he was very concerned with the current, universal brokenness of that relationship. This seemed enough for me, too.

Now I was fully immersed in biology courses. My favorites were comparative physiology, comparative anatomy and embryonic development. As a growing Christian, I was grateful that my professors didn't seem to be out to shipwreck my faith. In fact, they almost never addressed it. The closest was Jim Edwards, my comparative anatomy professor, who later went on to the National Science Foundation. He began the class by saying that although some might disagree, he would assume evolution was the best scientific explanation for the diversity of living things. Through many smelly dissections I learned firsthand about homology—the identification of the same anatomical elements in fish, amphibians, reptiles and mammals, a phenomenon well explained by evolution. The next year I was an undergraduate teaching assistant in the course while I was a student leader in the local Campus Crusade for Christ ministry.

My experiences in biology never generated debilitating cognitive dissonance. I never felt the need to tell Jim Edwards or other professors what they wanted to hear without actually engaging with their material. Perhaps I was helped by Lewis, or by the pastor of the Baptist church I attended, who encouraged us to hold firmly to Christian truths while pursuing excellence in our studies. I was too immersed in campus ministry to develop a comprehensive approach to thinking about evolutionary biology and Christian faith then. There would be time for that later.

I originally planned to pursue an MD/PhD, and in my senior year I received offers from very good schools to do so. It is a long, exciting story of God's calling, but after much prayer ultimately I decided to turn down those offers to pursue a Master of Divinity degree. I threw myself into biblical languages, biblical studies and theology. In those days there were battles afoot in evangelical seminaries. The biggest involved the nature of the Scriptures: how were they authoritative? What was inerrancy? Crucial questions to be sure, and yet my seminary curriculum, while strong on inerrancy, curiously lacked any discussion about science and the Bible, especially evolution. Perhaps this allowed me some flexibility. After all, one of the architects of the Chicago Statement on Biblical Inerrancy, J. I. Packer, had said that the two topics had little to say to one another. I emerged from seminary committed to biblical authority and scientific integrity, but resisting the temptation to force a premature agreement that might prove brittle and ill-advised.

While in seminary, I met the woman who would become my wife, Susie. She worked in the campus ministry of Campus Crusade for Christ, and had moved to the University of California-Berkeley. As I visited her during our courtship, I felt a strong call to pursue doctoral work in the sciences, and enrolled in the Biophysics PhD program at UC-Berkeley, where I developed my lifelong passion for understanding embryonic development.

Surely at Berkeley I would confront evolution head on! But in those days there were no sequenced genomes. For most bench scientists using model organisms (I studied sea urchin embryos then), it simply wasn't necessary to think much about evolution. Many—including me—thought that the amazing discoveries occurring in the fruit fly, *Drosophila*, were unlikely to carry over at a fundamental level to more "complicated" animals like humans. I would be wrong!

After Berkeley, Susie and I and our two boys moved to the Zoology Department at Duke University for my postdoctoral work. There I met Greg Wray, who was finishing his PhD just as I was arriving. Greg was the son of Methodist missionaries. He introduced me to a field now known as "Evo Devo" (short for Evolutionary Development). Greg and others were showing deep molecular similarities between widely diverse animals consistent with their common ancestry.

In 1991 we came to the University of Wisconsin, where I have been a faculty member in the Department of Zoology ever since. It has been a wonderful, blessed life. And yet I recognize that I am an endangered species. Elaine Howard Ecklund's survey, in which I was a respondent (reported in her book *Science vs. Religion: What Scientists Really Think*), places me in a tiny minority of my peers: evangelical scientists comprise only 4% of all science professionals at major research universities. Perhaps because of this I feel a special obligation to help my colleagues at the university and my brothers and sisters in Christ to see that there is no need to pit Christian faith against modern science, including evolution.

As a Christian asked to speak in churches and to student ministries on university campuses, I am committed to dynamic, personal faith in Christ and to God's Word, even amid disagreement. I emphasize that Christians can consider a variety of scientific options and remain faithful. One of my most poignant interactions involved a UW sophomore named Jeremy. Jeremy came to me one day with great curiosity. During a cell biology lecture he had noticed that my laptop's hard drive was named "Narnia," in honor of C. S. Lewis's children's novels. I explained that I was a Christian, and Jeremy proceeded to pour out his heart. Coming to UW from a very conservative Christian school, he felt forced to choose: he could either retain his Christian faith and jettison what he was learning about science, or accept science and abandon his faith. As we talked and read together over several months, Jeremy came to see that he could embrace the Lord of both biological life and life eternal. He eventually pursued a PhD in plant biology at the University of Texas at Austin.

It is not only my hard drive that is being watched, of course. At most public talks members of the local Skeptics group, who faithfully attend these events, carefully watch me for scientific inconsistencies. Apparently I've passed muster, since one of the Skeptics invited me to one of their Sunday public science events at a local pub! While I am glad to show integrity to skeptics, integrity before my colleagues is vastly more important. I want to show them that one can embrace modern science enthusiastically while maintaining robust Christian faith. There is a pragmatic issue, too. Biocore and Zoology teach evolution to thousands of undergraduates every year. As Chair, I owe my colleagues as much integrity regarding our teaching mission as I can muster. To me integrity of craft and heart are possible because I believe God has used evolutionary processes to bring forth his lavish creation. This view—evolutionary creation—remains faithful to God's Word, and to his world.

Back to 2009, when I found myself perspiring uncomfortably in a TV studio. There was nothing remarkable about my scientific presentation. I explained evolution and discussed some of the evidence from Darwin's day that makes the most sense if organisms share common ancestry (evidence like transitional forms in the fossil record, vestigial traits, island biogeography and homologies). I went on to explain that modern genetics provides even more impressive "molecular fossils" in our DNA: (1) genetic family trees based on DNA sequences, (2) proteins used to control embryonic development (like the fruit flies I maligned in graduate school!) that are retained in humans, (3) precisely shared patterns of "pseudogenes" (genes that have lost their functions over time) and (4) precise changes in human chromosomes versus those of other primates that are best explained by evolution. The list could go on.

Time had run out on my interview. There was only time for a last dangling question: How could I make sense of evolutionary biology in light of ancient faith? A good question—mercifully there was too little time left to provide a thorough answer! I give thanks that there are many Christians with deep faith in Jesus Christ, tenacious trust in the Bible as God's Word, and a strong commitment to celebrating the intricacies of the living world as we receive it from God's hand. They are thinking hard about this question. One important ingredient in any answer is a commitment to apply the right interpretive approaches to the book of God's Word and the "book" of his world. Evangelical scholars have been crucial here in helping the church read ancient documents as they were originally intended to be read. Second, I believe that while the church should passionately affirm each of these two "books," we must resist the temptation to insist on an excessively tight articulation between each, given our limited human understanding. Third, we need to provide a space where godly people can engage in edifying dialogue about difficult subjects. Convinced of this, I have edged toward more public, prayerful, winsome involvement through local initiatives to bridge the worlds of faith and science (for example, see isthmussociety.org) and through BioLogos. Much is at stake in these issues, and it is my prayer that no one on any side of these issues should act dismissively or hastily.

Challenges remain, of course. How should we make sense of the biblical account of Adam and Eve? How did sin enter the world and how is it transmitted? How does God's providence work through natural processes, including evolution? Whereas our theories about such things will likely require many revisions, there are realities beyond dispute. As G. K. Chesterton quipped so keenly, "Original sin . . . is the only part of Christian theology which can really be proved."[1] Or as Paul has it, "all have sinned and fall short of the glory of God" (Romans 3:23). Whatever our theories of the origins and transmission of sin, we are all—each one of us—in desperate need of a Savior. God's costly grace is available to each of us through Jesus Christ in his life, death and resurrection. Thanks be to God that he is indeed the Lord of life—now and forever.

7

Peace

Stephen Ashley Blake

Stephen Ashley Blake *is president of Realm Entertainment, a Los Angeles-based motion picture and television production company. His personal credits include feature films, television programs and music videos for studios, networks and record labels including Universal, Paramount, Warner Brothers, Sony, Fox, HBO, Geffen and Capitol Records.*

■ ▪ ■ ▪ ■

FOR MUCH OF MY LIFE I HELD NEW AGE views that sought to level the playing field among the world religions and synthesize them into a sort of common faith. Achieving this "equality" meant diminishing the extraordinary claims of certain religious figures that would otherwise make them stand out—particularly Jesus, whose role in the creation of the universe, power over life and death, and resurrection from the dead were downplayed or dispensed with altogether.

This was no problem for me, because although I had always respected Jesus, I chafed at the arrogance of Christians who claimed a special place for him in the religious pantheon. At the same time, though I held a certain vague respect for the Bible, there was always a foreboding about it, which ironically kept me from reading it for myself. Overall there was, in fact, a kind of mindlessness to my belief system. I had never actually worked out my views for myself or critically challenged them. Instead I had absorbed them from friends, family and the culture. In all of this, one thing was certain: you could never have told me I'd become a follower of Jesus.

That changed in 1995 when my sister Jaimie, a New Ager like me, unexpectedly became . . . a follower of Jesus. Not only was she passionate about her new faith—she had started praying at a workplace Bible study that her family would come to know Jesus too.

One night, Jaimie visited me with a pressing topic. Did I really understand the grave consequences of *sin*? I squirmed. Not only was the concept of "sin" utterly foreign to my thinking, but as a freewheeling bachelor, my life was, well, filled with it. After speaking on God's promised judgment for all sin (terrifying), she went on to his mercy and grace (comforting), and the forgiveness he offered through the sacrifice of the sinless Jesus, who in his execution took all punishment for humankind's sin upon himself. By accepting this freely extended gift of forgiveness, she had been "saved" from divine judgment and given eternal life. She was there to extend that gift to me. Like everyone else, I had always heard pithy phrases like "Jesus died for your sins," but no one had ever actually explained them to me. And there it all was, in that Bible I had so fearsomely respected. Jaimie gave me hers.

My first reaction was to reject thinking of myself as a "sinner." But I couldn't deny humanity's fallen condition—everything from my city to the world at large seemed tainted with some measure of wrongdoing—not to mention my own life. Then Gandhi came to mind. As a great respecter of his quest for right-eousness, I'd always marveled at his "frankest admission" of his "many sins" and failure "to reach the ideal which I know to be true." Even Gandhi was a sinner.

My resistance ran out of steam. Accepting the reality of my sins—and thus my need for forgiveness—and sensing God's love and care for me in Jesus' sacrifice, I became a believer that night. Within months, our entire family would come to believe.

At the time, my constant travels between Los Angeles and New York gave me ample time to read, and I began to devour the Scriptures book by book, chapter by chapter. I read my sister's pink study Bible in planes, cabs and subways, and on one occasion even abandoned an opera at the Met so that I could get back to my hotel and read. The more closely I followed Jesus, the more he filled my heart and life. Increasingly, the lost, empty, escapist life I had been living gave way to one of faith, security and self-control.

One of the most radical changes involved my work as a music video director. I had made a name for myself in gangsta rap, a genre whose powerful—if violently countercultural—lyrics captivated me and made for poignant narratives and imagery. But a train wreck was in the offing. Creating this dark content during the workweek and then showing up at church on Sunday to worship a holy God brought about an escalating crisis of conscience. It came to a head one day in prayer when I was overwhelmed by the conviction that my "double life" needed to end once and for all. I can only imagine what my clients thought of me when I told them why I would no longer be available for this work.

Because gangsta rap was my mainstay, I quickly went broke, and it would be years before I regained my financial footing. But I was ecstatic about my new life: I had a world of new friends and was filled with a peace I had never known. For the first time, my thoughts turned from "musical beds" to marriage. I also attended a strong Bible-believing church where the gospel was fearlessly preached—which, for all its tremendous upside, would bring about another crisis.

As a young evangelical growing in faith, I was not only embracing new spiritual perspectives, but a host of scientific ones. The universe wasn't ancient after all but in fact very young. Evolution was not only ridiculously false, but an atheistic attack on God, and to believe otherwise was to discredit the gospel itself. My church's anti-old-earth, anti-evolutionary stances were unapologetically militant, and because I had never critically challenged or investigated them for myself, I, like many others, adopted them with little reservation.

But maintaining these views within my world of experience brought unending conflict. All it took was watching a PBS science documentary, visiting an observatory or just overhearing scientific discussions among bystanders to remind me that my perspectives clashed with mainstream scientific thought.

I was, in effect, at war with science.

By then I had studied theology in depth and could discourse on any number of subjects, but the one area I had skirted—probably out of the same foreboding I had initially felt towards the Bible—was creation. *What processes did God use to create? Which version of science is accurate? Would the character of God permit him to reveal one reality in Scripture yet suggest a conflicting one in creation?* The issue had become unsettling, and the only way to resolve it would be to commit to a deeper investigation of Scripture and science, and far deeper prayer, so that I could reach my own informed conclusions.

My pillar of conviction going in was that God is not deceptive. Rightly interpreted, Scripture and science should form a cohesive—not clashing—portrait of Creator and creation. And as uncomfortable as it made me, I accepted that getting to the truth must mean allowing my presuppositions to be challenged—even dismantled—if they didn't hold up to scrutiny (Proverbs 18:17). So I resolved to study both the evolutionary and anti-evolutionary materials (scientific studies, biblical commentaries, formal debates and so on) with as much objectivity as possible, allowing each side to make its best case and refute the other. Thus my journey began.

As I delved into evolutionary theory, I was immediately struck by its eminently logical lines of reasoning, and caught off guard by the rational sense it made of the scientific data (my church had taught that macroevolution is an irrational, baseless "theory in crisis"). As I examined the perspectives of evangelical theologians and scientists who found no conflict between Scripture and evolution—many of whom were biblical inerrantists—mainstream science gained great credibility in my thinking.

In the young-earth creation materials, I was, as always, impressed by the high regard for Scripture. But I was disillusioned to discover that their scientific assertions seemed at blatant odds with natural reality—in fact, the young-earth perspectives seemed to have been written in a sort of empirical vacuum, as if the products of wishful thinking rather than experimentation. While I had hoped to find young-earth creation science viable in at least some of its applications, I was dismayed to discover just how thoroughly unworkable it was. Not only did nothing about a young universe seem to add up, but as far as I could tell, such theories were so densely loaded with scientific absurdities and impossibilities as to virtually certify them as false. When I contacted a high-level representative at a very prominent young-earth creationist organization, he acknowledged the "apparent" unworkability of young-earth cosmology before candidly sharing his way of dealing with it: rather than focus on the science, he let his longstanding interpretation of Genesis—which he said was not open to reconsideration—wholly determine his stance.

Alarm bells went off in my head. Unwillingness to critically challenge one's interpretative methods sounds dangerously close to declaring personal infallibility.

This brought me to a crisis: If evolutionary science was truly fallacious, as we evangelicals are regularly taught, why does it lead to such astonishing breakthroughs, while "accurate" young-earth science remains so completely barren and unworkable in practice? No one can deny mainstream science's spectacular, ever-growing achievements in everything from space exploration to genetics. Is God deceiving humankind by prospering a "heretical" (the actual term my pastor used) view of creation while making "true" science look resoundingly false?

One of the issues that disturbed me most about evolution was the randomness it required. The concepts of divine sovereignty and "purposeless chance" seemed irreconcilable. But I came to realize that with respect to both the structure of the universe and our own lives, "random" occurrences at the micro level are in fact the constituents of order and stability at the macro level, and that God uses events that we observers legitimately perceive as "chance" to bring about his intended ends. In fact, it struck me that Christians in particular should be able to relate to this apparent dichotomy. Despite at times experiencing apparently random events in our own lives, we steadfastly affirm God's sovereignty over all.

Poised to accept evolution on scientific terms but still wrestling with purported contradictions with Scripture, I returned to the biblical creation accounts. I had long been taught that anything but a literal interpretation of these texts would discredit the Bible, but I was now surprised to find its figurative imagery leaping off the page—God was masterfully using evocative imagery to convey eternal truths. I began to see a beauty in Genesis that I had never before known. For example, I found

Genesis 2:7—which depicts God fashioning his precious, beloved children from the "nothing" of dust—not only humbling, but profoundly elegant and consistent with evolution.

The war had finally ended. I had come to peace with science. Today, I strongly support the endeavor of modern science, and view its practitioners—whether believers or not—as revealing our Creator's wondrous ways to humankind. Unfortunately, the very act of holding the young-earth view up to scrutiny caused conflict with church leadership, and was a factor in my departing that church.

I am deeply troubled by many of my fellow evangelicals' hostility towards science, as I believe it threatens to destroy the faith of our children and grandchildren—who today are vehemently taught that evolutionary theory is incompatible with the cross of Christ, and tomorrow will enroll in universities that demonstrate its manifest integrity, setting the stage for cataclysmic crises of faith for countless young believers.

Flash forward. My wife, children and I now attend a church that is theologically solid and focused on kingdom activity. Our pastoral staff recognizes the church's responsibility to wrestle through these crucial issues, and the need for believers to do the same. It has also been our privilege to work with local churches in hosting faith-science discussion events in which believers and skeptics alike can explore the tremendous things of Christ, through whom and for whom all things in heaven and earth were created.

- 8 -

Learning the Language of God

Francis S. Collins

Francis S. Collins *is a physician-geneticist. He led the Human Genome Project, which produced the first reference sequence of the human DNA instruction book in 2003. He is director of the National Institutes of Health, founder of BioLogos, a guitar player and Harley rider, and author of* The Language of God.

■ · ■ · ■

FAITH WAS NOT AN IMPORTANT PART of my childhood. I was vaguely aware of the concept of God, but my own interactions with him were limited to occasional childish moments of bargaining about something that I really wanted him to do for me.

After graduation from college, I went on to a PhD program in physical chemistry at Yale. I gradually shifted from agnosticism to atheism. I felt quite comfortable challenging the spiritual beliefs of anyone who mentioned them in my presence, and discounted such perspectives as sentimentality and outmoded superstition.

Two years into this PhD program my narrowly structured life

plan began to come apart. Despite the daily pleasures of pursuing my dissertation research on theoretical quantum mechanics, I began to doubt whether this would be a life-sustaining pathway for me. In an effort to broaden my horizons, I signed up for a course in biochemistry. The ability to apply rigorous intellectual principles to understanding biology, something I had assumed impossible, was bursting forth with the revelation of the genetic code. I was astounded.

Though I was well along with my PhD program, and after much soul searching, I applied for admission to medical school. I was accepted at the University of North Carolina. Within a few weeks I knew medical school was the right place for me. As physicians in training, medical students are thrust into some of the most intimate relationships imaginable. What struck me profoundly about my bedside conversations with these good North Carolina people was the spiritual aspect of what many of them were going through. My most awkward moment came when an older woman, suffering from severe untreatable angina, asked me what I believed. I felt my face flush as I stammered out the words, "I'm not really sure."

The moment haunted me for several days. I had been confident that a full investigation of the rational basis for faith would deny the merits of belief, and reaffirm my atheism—but I had never really looked. Now I was determined to do so, no matter what the outcome. In the course of my survey through the major religions of the world, a Methodist minister who lived down the street took a small book off his shelf and suggested I read it. The book was *Mere Christianity* by C. S. Lewis. In the next few days, I realized that all of my own constructs against the plausibility of faith were those of a schoolboy.

I had started this journey of intellectual exploration to confirm my atheism. Agnosticism, which had seemed like a safe second-place haven, now loomed like the great cop-out it often is. Faith in God now seemed more rational than disbelief. I was beginning to understand from looking into my own heart that the evidence of God's existence would have to come not from science, but from other directions, and the ultimate decision would have to be based on faith, not proof. For a long time I stood trembling on the edge of this yawning gap. Finally, seeing no escape, I leapt.

But that was the beginning of the journey to find the truth, not the end. I still needed to consider the claims of faith—and most especially the most outrageous of all those claims, that Jesus Christ was not just a wise teacher, but the Son of God, and that he literally rose from the dead. The historical evidence was compelling, but could I believe? A full year had passed since I decided to believe in some sort of God, and now I was being called to account. On a beautiful fall day, as I was hiking in the Cascade Mountains, the majesty and beauty of God's creation overwhelmed my resistance. As I rounded a corner and saw a beautiful and unexpected frozen waterfall, hundreds of feet high, I knew the search was over. The next morning, I knelt in the dewy grass as the sun rose and surrendered to Jesus Christ.

Several years ago, I spoke to a national gathering of Christian physicians. I explained how I had found great joy in being both a scientist studying the genome and a follower of Christ. Warm smiles abounded; there was even an occasional "Amen." But then I mentioned how overwhelming the scientific evidence for evolution is, and suggested that in my view evolution might have been God's elegant plan for creating humankind. The warmth left the room. So did some of the attendees, literally walking out, shaking their heads in dismay.

What's going on here? From a biologist's perspective, the evidence in favor of evolution is utterly compelling. Darwin's theory of natural selection provides a fundamental framework for understanding the relationships of all living things. The predictions of evolution have been borne out in more ways than Darwin could have possibly imagined when he proposed his theory 150 years ago, especially in the field of genomics.

If evolution is so overwhelmingly supported by scientific evidence, then what are we to make of the lack of public support for its conclusions?

The problem for many believers is that the conclusions of evolution appear to contradict certain sacred texts that describe God's role in the creation of the universe, the earth, all living things and ourselves. Evolution raises questions. These questions have been debated for centuries. Nonliteral interpretations since Darwin are somewhat suspect in some circles, since they could be accused of "caving in" to evolutionary theory, and perhaps thereby compromising the truth of the sacred text. So it is useful to discover how learned theologians interpreted Genesis 1 and 2 long before Darwin appeared on the scene, or even before geologic evidence of the extreme age of the earth began to accumulate. Upon examination we find that Christian theologians have affirmed diverse interpretations of the meaning of Genesis 1 and 2.

It is true that some, particularly from the evangelical Christian church, insist upon a completely literal interpretation of Genesis, including twenty-four-hour days. Coupled with subsequent genealogical information in the Old Testament, this leads to Bishop Ussher's famous conclusion that God created heaven and earth in 4004 BC. Other equally sincere believers do not accept the requirement that the days of creation need to be twenty-four hours in length. But they otherwise accept the narrative as a literal and sequential depiction of God's creative acts. Still other believers see the language of Genesis 1 and 2 as intended to instruct readers of Moses' time about God's character, and not an attempt to teach scientific facts about the specifics of creation that would have been utterly confusing at the time.

Despite twenty-five centuries of debate, it is fair to say that no human knows what the meaning of Genesis 1 and 2 was precisely intended to be. We should continue to explore that! But the idea that scientific revelations would represent an enemy in that pursuit is ill conceived. If God created the universe and the laws that govern it, and if he endowed human beings with intellectual abilities to discern its workings, would he want us to disregard those abilities? Would he be diminished or threatened by what we are discovering about his creation?

The relatedness of all species through the mechanism of evolution is such a profound foundation for the understanding of all biology that it is difficult to imagine how one could study life without it. Evolution, as a mechanism, can be and must be true. But that says nothing about the nature of its author.

Evolutionary creation is the dominant position of working biologists who are also serious believers. That includes Asa Gray, Darwin's chief advocate in the United States, and Theodosius Dobzhansky, the twentieth-century architect of evolutionary thinking.

Here is a synopsis of the position: God, who is not limited in space or time, created the universe and established natural laws that govern it. Seeking to populate this otherwise sterile universe with living creatures, God chose the elegant mechanism of evolution to create microbes, plants and animals of all sorts. Most remarkably, God intentionally chose the same mechanism to give rise to special creatures who would have intelligence, a knowledge of right and wrong, free will and a desire to seek fellowship with him.

This view is entirely compatible with everything that science teaches us about the natural world. It is also entirely compatible with biblical Christianity. The evolutionary creation perspective cannot, of course, prove that God is real. No logical argument can fully achieve that. Belief in God will always require faith. But this synthesis has provided for legions of scientist-believers a satisfying, consistent, enriching perspective that allows both the scientific and spiritual worldviews to coexist happily within us. This perspective makes it possible for the scientist-believer to be intellectually fulfilled and spiritually alive, both worshiping God and using the tools of science to uncover some of the awesome mysteries of his creation.

Believers should seek to be in the forefront among those chasing after new knowledge. Believers have led science at many times in the past. Yet all too often today, scientists are uneasy about admitting their spiritual views. To add to the problem, church leaders often seem to be out of step with new scientific findings, and attack scientific perspectives without fully understanding the facts. The consequence can bring ridicule to the church, driving sincere seekers away from God instead of into his arms. Proverbs 19:2 warns against this kind of well-intentioned but misinformed religious fervor: "Zeal is not good without knowledge" (HCSB).

It is time to call a truce in the escalating war between science and the Spirit. The war was never really necessary. Like so many earthly wars, this one has been initiated and intensified by extremists on both sides, sounding alarms that predict imminent ruin unless the other side is vanquished. Science is not threatened by God—it is enhanced. God is most certainly not threatened by science; he made it all possible.

So let us together seek to reclaim the solid ground of an intellectually and spiritually satisfying synthesis of all great truths. That ancient motherland of reason and worship was never in danger of crumbling. It never will be. It beckons all sincere seekers of truth to come and take up residence there. Answer that call. Abandon the battlements. Our hopes, joys and the future of our world depend on it.

9

Faith, Truth and Mystery

Oliver D. Crisp

Oliver D. Crisp *is professor of systematic theology at Fuller Theological Seminary. He has written and edited a number of books, including* Retrieving Doctrine: Essays in Reformed Theology *(IVP Academic). He is married and has three children, a Basset hound called Watson and a cat called Mycroft.*

■ ▪ ■ ▪ ■

THREE THEOLOGICAL PRINCIPLES HAVE BEEN important to me as I have reflected on how faith connects with evolution. The first of these I hit upon when I began to study the work of St. Anselm of Canterbury many years ago: *faith seeking understanding.* We begin from a position of faith, and seek to understand what we are committed to through intellectual reflection on that faith—a rationale that has always struck me as profoundly biblical (compare Hebrews 11:6). For me that has meant seeking to understand the faith I have been given using the tools provided by Christian theology and analytic philosophy, since those are the academic disciplines in which I have been trained.

A second principle is: *all truth is God's truth.* I am not entirely sure where I first came across this principle, though I think it may have been through the work of the Christian philosopher Arthur Holmes, whom I first read as an undergraduate at the University of Aberdeen. The idea behind this principle is a sort of realism about truth. There are things that are true independent of what we think about the matter. For instance, it is true that the earth revolves around the sun irrespective of whether or not I believe it. One theological consequence of this principle is that, if a thing is true, it is true irrespective of where one happens upon it, because God has created the world in such a fashion that it reflects these truths, which we are able (at least in some measure) to access. Where they are not revealed in Scripture, they are there to be discovered in the world around us. So when the Christian comes to read the Bible, she believes that it is a means of God's speaking. He reveals himself and his purposes for us and for our salvation there in a way that he does not reveal himself in, say, Shakespeare's plays. Nevertheless, the truth that we find in Shakespeare's plays is true to the extent that it measures up to the truth that God has woven into the world, including, of course, the human mind. The same is true of the world around us. There are certain physical constants in the world, and we can know them through observation and calculation. There is a truth to be had about what these constants are, due to the fact that God created the world this way. He has created it to be a place where these constants are in evidence, and where we are able to uncover them by means of observation and calculation. This is summed up in the hymn "This Is My Father's World."

This is my Father's world, the birds their carols raise,
The morning light, the lily white, declare their
        Maker's praise.

This is my Father's world: He shines in all that's fair;
In the rustling grass I hear Him pass; He speaks to me
        everywhere.

The third principle is that *God is mysterious.* This is easily traced back to Scripture. For instance, in the Old Testament we are told that the secret things belong to the Lord our God (Deuteronomy 29:29). Moreover, Job is asked, "Can you fathom the mysteries of God? Can you probe the limits of the Almighty?" (Job 11:7). The apostle Paul says in the opening chapters of 1 Corinthians that he declares God's wisdom in the gospel, not the wisdom of his own age but "a mystery that has been hidden and that God destined for our glory before time began" (1 Corinthians 2:7). Many other biblical passages tell a similar tale. God is beyond our ken. Yet he reveals something of himself in Scripture, in Christ, and even in the world around us for those who, with the eyes of faith and the spectacles of Scripture, are able to see it.

These days *mystery* has become something of a pejorative term in some circles. It is commonly thought that theologians cry "mystery!" when they are unwilling to accept the implications of the views to which they have committed themselves. Although there are certainly unprincipled uses of this theological category in the history of Christian thought, the misuse of a thing doesn't mean there isn't a right use. If there is a God, he is bound to be mysterious because he is so vastly different from us, so unlike what we are. In his Victorian romance *Flatland*, Edwin A. Abbott envisions a world of two-dimensional creatures. What would it be like for such a creature to begin to dimly grasp the notion that there might be three-dimensional beings? It would be a colossal imaginative leap. Just so with our grasp of God, who is infinitely greater than mere mortals. Given that there are many mysteries we cannot fathom independent of any theological commitments (for example, human consciousness, or how the physics of Einstein can be reconciled with the physics of the quantum world), it cannot be that the notion of mystery itself is the problem. It must be the use to which it is sometimes put. Yet there is a right and proper use of this notion in theology as there is in other areas of human endeavor, such as the philosophy of mind.

These are hardly the only three theological principles that have formed me in important respects, and they alone are insufficient to yield a particular way of thinking about the relationship between evolution and Christian faith. I suppose that St. Anselm held to all of these principles, though he knew nothing about Darwinian evolution. But I think these principles are salient because they provide important reasons for taking seriously the project of the natural sciences as we now know them, as well as reason for thinking that Christians have nothing to fear from such a project. Let me explain.

I presume that God is essentially good because he is a perfect being ("As for God, his way is perfect," Psalm 18:30). I also presume that he loves his creatures and desires the best for them. Thus it is not implausible to presume that the world he creates will be consonant with his character. It will reflect his goodness and his lovingkindness towards what he has created. This is what we find reflected in the books of Scripture. But it is also what many Christians have claimed that we find in the "book" of the world around us. In other words, the character of God revealed to us in holy writ must match up to the character of God we can discern in the created order if God is essentially good and loving. He does not deceive us. He does not seek to mislead us. He seeks our happiness and our good.

Yet since the nineteenth century, natural scientists have told us that the world is a place that has evolved over millennia, and that this has occurred in the biological world according to a process of natural selection, which privileges those creatures able to pass on biological traits that are advantageous in surviving changing circumstances and environmental conditions (among other things).

It is no wonder that many Christians find this difficult to square with their religious faith, and end up fleeing from evolutionary accounts of the world to the safe haven of a premodern worldview. It is still the case in many parts of the evangelical subculture today that one must choose between belief in the God of the Bible and belief in the godless world of Darwin and his heirs. The two things are often thought to be simply incompatible.

And yet if evolution is true, and all truth is God's truth, and faith seeks understanding—of the mysterious God revealed to us in Scripture, in Christ and (in some dim sense, perhaps) in the world around us—then evolution and biblical Christianity must in principle be consonant with one another even if it sometimes appears that they are in conflict from our particular vantage point. This is the approach that has informed my own thinking about these matters.

Recently amongst the chattering masses on the Internet there was a much-talked-about picture of a woman's dress: "that dress," as it came to be known. A picture circulated which some perceived to be of a gold and white dress while others, *looking at exactly the same photograph*, saw a black and blue dress. How can this be explained? It turns out that the riddle had a perfectly reasonable answer, having to do largely with how we perceive objects in different light. When I came across this discussion it immediately made me think of the creation-evolution debate in contemporary evangelical Christianity. Two people with the same Christian worldview, the same evangelical views about the nature of Scripture, how the world was created by God, and so on, can yet come to different views on the relationship of evolution to creation. It is too simplistic to reduce these differences to mere perception. But there is something about this debate that is like the issue of "that dress." Those in possession of the same fundamental Christian beliefs can form different views about the relationship between evolution and the created world.

Years ago when I was an undergraduate studying divinity in Scotland, I thought that no serious Christian could hold to the "theory of evolution," as we called it. It was simply incompatible with Christian faith, because it is a metaphysically naturalistic explanation of the world around us. That is, it proposes that the world evolved via various physical processes that include natural selection, which is undirected by its very nature—it has no place for God. However, in reflecting on the evidence and the arguments of many in the scientific community in the intervening years, as well as thinking about the philosophical and theological issues these raise, I have come to see that this was a mistaken view. My faith has not changed; but my understanding of that faith has certainly developed. My view of God has not diminished, for if evolution and the Bible are both true, then that is surely because all truth is God's truth. There may be superficial differences between the two, yet deep consonance, as the Christian philosopher Alvin Plantinga has recently reminded us in his book *Where the Conflict Really Lies*. Finally, if God is a mysterious being that has created and sustains all things, is it any wonder that the details of how these two things are related are not always as clear to us as we would like? Isn't that true of many things in the world around us? And isn't that what we would expect if there is a God like the one Scripture proclaims, who is mysterious, whose ways are past finding out, and whose purposes are not always shared with his creatures? (A fact made abundantly clear to Job when he encountered God!)

I do not presume to have all the answers to important and pressing theological questions raised by evolution as it is understood by those working in the natural sciences today. But I think I see the connection between the two a little more clearly than previously. I presume that God ordains natural processes, including natural selection. Many of religion's cultured despisers think this naive in the extreme, since natural selection is supposed to be unguided. But I don't see why the Christian must presume this; such a claim is not a deliverance of science as such, but a statement of metaphysics. Many infer that there is no intelligence guiding what happens in the world. However, it is perfectly reasonable for a Christian to appeal to both the deliverances of science as science and the deliverances of Scripture as revelation. For we presume that Scripture is a source of knowledge as well. And if all truth is God's truth, then *in principle* our understanding of Scripture and science are compatible, even if the precise manner in which they are compatible may not always be clear to us—as I have already indicated is the case with many other such conundrums in human intellectual endeavor.

10

Inspired by an Amazing Universe

Jennifer Wiseman

Jennifer Wiseman *is an astronomer, speaker and writer. She received her PhD in astronomy from Harvard University and studies the formation of stars and planets in interstellar clouds using radio, optical and infrared telescopes. She is also interested in public science policy and outreach, and enjoys giving talks about the excitement of astronomy and scientific discovery. She loves animals and the beauty of the natural world.*

■ ▪ ■ ▪ ■

I GREW UP ON A FARM IN A RURAL PART of the Ozarks in Arkansas. We were surrounded by nature all the time: plants, animals, livestock, pets and a nice view of the night sky. Many evenings my parents and I would go for long walks down the country lane where we could see countless stars, from horizon to horizon, and it made me wonder what was out there. I came of age as NASA was sending the first probes to other planets in our solar system. These probes sent back images of exotic worlds like Europa and Io, two of Jupiter's moons. I wanted to go and explore, either as an astronaut or at least with telescopes, viewing from a distance. My early exposure to nature played a big part in my future pursuit of astronomical science as a career.

My childhood home was also a loving Christian one, and I am very grateful for that. I experienced the love of Christ both in my biological family and in our church family. We did not talk much about science in church, but we did honor God as the author of all creation. We sang hymns that I still love today, like "This Is My Father's World." As I grew up, I never saw any conflict between science and our faith. We understood, perhaps in a simplistic way, that God was responsible for nature, and because nature was magnificent, God was to be praised for the natural world.

I did not know exactly how to proceed toward a scientific career, but I was excited to be able to start down that path by attending a university. With encouragement from my older, science-minded brother, I was the first person in my family to earn a science degree from college. As an undergraduate student at MIT, I realized that there was potentially some conflict between what I was learning in my science classrooms, or hearing in the offhand remarks of professors or in the media, and what my faith affirmed. I realized it had little to do with the science, and more with how we interpret Scripture. In my home church most of us accepted a fairly literal interpretation of Genesis. We did not have any reason to interpret it any other way. We understood that God created everything in six days, as outlined in Genesis. But that was always taught with humility. Our pastors said things like, "Don't be too presumptuous about these things; God may not have revealed all the details of how he created in these few verses in Scripture." We understood that we needed to be open to the fact that with God a day could be like a thousand years, and he may have used great ages of time and great care in creation.

It was that foundation of humility that prepared me to reconcile what I was learning in my college classrooms with Scripture.

My scientific education did not cause me to doubt what the Bible says regarding God's authoritative involvement in creation, but I started reading a lot more, particularly books by scientists who were Christians about how they reconciled their understanding of Scripture with what they had learned scientifically about the details of nature. Some of these scientists also came to our Christian fellowship groups on campus, and it was a terrific help to me as a student to see these models of excellent scientists who were followers of Jesus Christ. I saw their reverence for Scripture and God along with their love for studying the natural world fitting together in a beautiful mosaic.

As an astronomer, I have spent my career studying the heavens. Astronomy and cosmology reveal grand changes in the universe over billions of years. Since our telescopes capture light and other electromagnetic radiation that has traveled through time, we can actually observe earlier states of the universe. We can see "infant" galaxies as they were just forming not long after the dawn of creation, and study their compositions. We find that the stars in those early galaxies have very simple compositions— mostly hydrogen gas and not much else. But studying more mature galaxies like our own Milky Way, we find that after several generations of stars come and go over billions of years, these stars, acting as cosmic furnaces, have produced and dispersed heavier elements like oxygen, carbon and iron—elements we need and enjoy in our own time and solar system for planets and life. In other words, stars themselves are magnificent factories that have played a key role in the maturing and enrichment of the universe. Seeing this drama unfold on the cosmic scale of time and space for the whole universe makes it easier for me to see how biological evolution on planet Earth could fit into that grander story. It also infuses my own thinking about creation with a sense of both awe and humility.

Even though I don't experience an ultimate conflict between science and my Christian faith, I still have many unanswered questions. And when I consider the vastness of space and time, I am confronted with profound mysteries and a sense of God's greatness. The universe is over thirteen billion years old. But our earth has existed for only about four billion years, and higher life forms have existed for only a few hundred million years—or less, depending on how you define it. So what was God *doing* in all those ages before these familiar parts of our world existed? Was God just as interested in the welfare of the universe eight billion years ago, when there were galaxies but no planets? Or millions of years ago, when there were dinosaurs but no large mammals to speak of? Was God just waiting for humans to come around? Why didn't he create all of this instantly and just to get to the point? I don't know the answers to these questions. I suspect, though, that God views and uses time very differently than we do. Even human life spans do not seem very efficient, given that we sleep a large fraction of our lives. Our spiritual lives do not mature as quickly as we might hope. It takes a lifetime for us to mature in many different aspects. God seems to take an interest in the present, and also to have a sense of patience and a vision for what is going to eventually emerge, whether in the natural world or in our own personal lives. In Genesis, we read that God declared each stage of creation "good" in its own right. Thus I believe God has deeply cared about each phase of the universe since the beginning, even before humans were around to recognize him, and that each phase brings him honor. We may never know this side of heaven *why* he creates gradually. Perhaps in his mode of creation God is modeling the kind of patience required of us.

When some people hear that the universe has been around for billions of years, they despair and feel that life must be insignificant and pointless. In a universe that is expanding—a universe of billions of galaxies where stars come and go, and presumably their planets as well—how could our lives be significant? This is a good question that I can't easily answer, but I take heart in the fact that the Bible affirms over and over that we *are* significant. Not because of our lifespans or our position in the cosmos. In fact, Scripture quite explicitly says that we are like dust, and like the grass that comes and goes (Psalm 103:14; 1 Peter 1:24). And yet, we are also told over and over that God deeply loves us, and that we are made in God's image. Our significance is not because of the tiny space and time we occupy in the universe. It is because we are here at all, and because we can have a relationship with the God who is responsible for it all, and because this relationship can carry on eternally. Scripture reveals profoundly that because God so loved the world, "the Word became flesh and made his dwelling among us" (John 1:14).

This way of thinking should greatly enrich our prayer life. The God we pray to is a God who is responsible for this universe that has been developing over unfathomable ages. Yet, before the creation of the world, he had us in mind. He cared about us and had a plan for our lives and redemption. He planned to become part of creation in Jesus Christ. This God who is so mighty and loving and wise hears our prayers. That is amazing to consider!

The way we view God and our relationship to him is also enriched by considering the possibility of life on other planets. I have had the privilege of working with several major telescopes and observatories. Large telescopes have lately revealed the existence of many planets outside our solar system. In fact, we are discovering some planets that may be similar to Earth: they are a similar distance away from their parent star, they may have the ability to harbor liquid water, and they may have the conditions we think are conducive for life. It is an exciting time! We have not yet discovered life on planets beyond Earth, but the potential is there as we rapidly develop technology to investigate more details about these so-called "exoplanets."

What does that mean for Christians? Could there be life elsewhere? Again, we should approach these questions from a foundation of awe and humility. The Bible does not speak to this issue specifically. Christian theologians for many centuries have contemplated the implications of the possibility of life in other worlds and what that would mean for their relationship with God and the redemptive work of Christ. This is not a new topic, but the discoveries we are making are new, and they reinvigorate questions about the implications of the possibility of finding life beyond Earth.

As technology keeps advancing, we have the thrilling opportunity to discover more of the extravagance of God's creation. And as we search for the possibilities of life elsewhere, we can ask how finding such life might affect a Christian understanding of the relationship of God with humanity. To me, whether life exists elsewhere or we are all alone, we have reason to praise God: our faith does not hinge on what we discover through science. A universe filled with life would be perfectly in line with a very generous God who we know has created billions of galaxies and billions of stars within each galaxy, stars which themselves produce elements needed for life over great spans of time. On the other hand, our being alone in the universe (though we could never prove that scientifically without observing every possible star system!) would bring forth extreme gratitude that we're here at all. Personally, I would not be surprised if there is at least simple life beyond Earth, given that we see all kinds of interesting life forms in every environment on planet Earth. Why could not our generous and creative God, who enabled life and the conditions for life here on planet Earth, do the same in other places as well?

In closing, I'd like to offer some practical advice to church leaders who want to start a discussion about science in general or the evolutionary history of our world and universe in particular. Start with praise. Scripture is filled not with discussions of scientific details, but with praise—praise to God for his work in creation. The Bible proclaims that even the natural processes of the heavens and of animals and plants living out their lives are acts of praise. We ought to begin by being amazed at what is going on in the natural world and what we can learn from science. Let us be grateful for the tools of science and think of them as gifts from God that enable us to see the details of his handiwork.

Science will raise difficult issues. It will help reduce anxiety if we introduce those issues with genuine questions and a sense of humility. We may not find the answers for all these questions immediately. We may have to remain patient while we look at several different approaches that Christians have taken on such topics. And we may have to do the hard work of learning what is actually being discovered in science before we pass judgment on it. In this way humility can be built on a foundation of wonder, knowledge and curiosity about the natural world. That leads us to realize it is okay to ask questions. Nothing we investigate in science is going to threaten the ultimate authority and the love of our good and faithful God.

11

Boiling Kettles and Remodeled Apes

John Ortberg

John Ortberg *is an author, speaker and the senior pastor of Menlo Park Presbyterian Church in the San Francisco Bay Area. His most recent book is* All the Places to Go . . . How Will You Know? *Now that their children are grown, John and his wife, Nancy, enjoy surfing the Pacific to help care for their souls.*

■ ▪ ■ ▪ ■

DOES SCIENCE DISPROVE OUR FAITH? We might start thinking about this by considering the question of whether science is the only reliable way to acquire knowledge. Science has great prestige in our day, so this is a really important question. Are there any other kinds of knowledge besides scientific knowledge? The short answer is yes, and if we don't recognize that, it limits the knowledge we have to live by. Because science has made such amazing progress in certain fields like medicine and technology, some people claim that the scientific method, or empirical verification, is the only way to reliable knowledge. That would mean there is no such thing as moral, spiritual or

This essay was adapted from John Ortberg's sermon "Does Science Disprove Faith?," given on January 5, 2014, at Menlo Park Presbyterian Church.

personal knowledge. This view that the scientific method is the only reliable way to knowledge is sometimes called scientism.

Sir John Polkinghorne is a Cambridge physicist and an Anglican priest, and he may be the greatest thinker about faith/ science issues in our day. He has a really helpful illustration. He invites us to imagine somebody asking, "Why is water boiling in that kettle?" One person answers, "Because burning gas is heating the water," and another person answers, "Because I want a cup of tea." Which answer is right? Well, they're both right. One person is talking about impersonal, mechanical causes. That's what science tends to do. The other answer is framed in terms of a person and purpose and intention. It is not scientific in a mechanistic way, but it's true and it's terribly important. So science involves a method that is enormously useful to investigate a large part of reality, but it is not the only way to know truth. For example, human life is of great value. That's true. You know that, but you can't put it in a test tube. It is wrong to live for selfish greed. That is moral truth. Scientism is a dogma that says any dimension that cannot be exhaustively explained by the scientific method doesn't exist or doesn't matter. Is science the only way to know something reliably? No, it's not. It's very important, but it's not the only way.

Sometimes we're told that science has shown that the universe has no purpose: that it's just a random machine. The late Carl Sagan said, "We find that we live on an insignificant planet of a humdrum star lost between two spiral arms in the outskirts of a galaxy which is a member of a sparse cluster of galaxies, tucked away in some forgotten corner of a universe in which there are far more galaxies than people."[1]

Notice the value-laden words: insignificant, humdrum, lost, tucked away, forgotten. Those are not scientific terms, but they're weighted with meaning. The idea behind statements like this is that somehow science, by showing us how immense the size and the age of the universe is, has shown us that little tiny human beings do not have unique dignity in ways that faith has taught. However, Sagan did not invent the idea that there's a contrast between the immensity of nature and the tininess and brevity of human life. The psalmist said thousands of years ago, "When I consider your heavens, the work of your fingers, the moon and the stars, which you have set in place, what is mankind that you are mindful of them, human beings that you care for them?" (Psalm 8:3-4). Precisely the same contrast was the object of serious reflection a long time ago. But the psalmist does not go on to say, "The way to settle this one is to look at the scale of stuff in terms of physical size. I think people are huge. I think the earth is huge and the sun and the moon and the stars are tiny, so we win." That's not what the psalmist says. The psalmist says, "Yet, God, you have created human beings with glory and honor. You've crowned them, made them something like transcendent beings." Human beings are invested with a divine image. They have this capacity to learn and create. They have the weight that comes with being a moral agent, being able to make decisions and be responsible for them, and being able to care for creation. It's staggering.

The nonhuman elements of the universe also evoke a sense of wonder in us that is remarkably stubborn. Last Christmas break, when I was surfing at Huntington Beach, I saw something I'd never seen before. Dolphins kept swimming by a few feet away from where I was, and a couple of them had a little calf. Just seeing a little dolphin calf in the ocean was quite extraordinary, but that wasn't the best part. At one point I looked up, and there was a big wave coming in (relative to my surfing skills), and at the top of it was a silhouette of a dolphin. I had never seen anything like this. You see posters where a bunch of dolphins have been photoshopped in, but this was the real deal—a dolphin on a wave. Then the wave broke, and the dolphin turned parallel to the shore and bodysurfed that whole wave while I just watched it go by. Then it popped out and said, "Hi, John." (No, it didn't do that, but it did the rest of it.) I'll never forget that. It was phenomenal.

Wonder is the indistinguishable realization not just that something is, but that it is good. It's the human heart echoing those words from Genesis, that God spoke and it was so, and God saw that it was good. And it is. We know that. Wonder moves us dangerously close to worship. If you're thoughtful, you have to ask, "Is our hunger for wonder and meaning a clue to something beyond material reality?" C. S. Lewis writes, "Creatures are not born with desires unless satisfaction for those desires exists. A baby feels hunger: well, there is such a thing as food. A duckling wants to swim: well, there is such a thing as water. [People] feel sexual desire: well, there is such a thing as sex. If I find in myself a desire which no experience in this world can satisfy, the most probable explanation is that I was made for another world."[2]

But now wait a minute; isn't it now clear from science that we're part of this world? Just the products of evolution? This has become a hot-button topic. Evolution is a really controversial thing. A little boy comes to his dad and asks him, "Dad, where did human beings come from?" and his father says, "Well, we descended from apes." The little boy goes to his mom. "Mom, where did human beings come from?" She says, "We were created by God in God's image." The boy says, "But Dad said we descended from apes." Mom says, "Well, I was talking about my side of the family." Funny, but is there something here similar to the tea kettle example? Does a scientific description conflict with a theological statement? Only if we understand the Bible as offering a competing scientific description.

Wheaton College Old Testament professor John Walton persuades me that the books of the Bible always emerged out of a conversation in their day. People get all kinds of goofy ideas about them if they assume, "I don't have to begin by looking at the historical context and asking what the initial readers would have understood this to mean. I can just read into it whatever I would happen to read into it out of my own time and culture and agenda." There's a kind of arrogance in that mindset: a failure in humility. Walton has spent a lot of time taking a look at this. There was a conversation in the ancient Mesopotamian world: "Where did we come from? How did the earth get here?" But it was very different from the conversation or the agenda in our day, and it's critical for how we understand Genesis.

I grew up in the church, and I didn't know about that ancient conversation. I just assumed the Bible was a magic book and that Genesis had dropped down out of heaven. It was threatening to me to find out there was a very rich conversation going on, and of course the language and concepts of that conversation were used by God in inspiring the writer of Genesis.

I believe the best reading of it, just on biblical terms, is that it's not about how God created, or how long it took, nor is it about the role of mutation or natural selection. Those questions were not around back then. Genesis lays out the identity of human beings and our place in the cosmos with matchless, world-changing truth. Thus it's legitimate for science to explore the questions about how and how long.

I have seen too many young people in too many churches exposed to bad science, shoddy thinking, false claims and misguided ideas (maybe well-intended but still misguided). It's easy to think we're defending the Bible when we're really defending a wrong interpretation of the Bible. Then I've watched when very often those really bright young people go off and pursue education, begin to read and discover they were misinformed. Then they think they have to choose between the Bible and truth. But they don't.

On the other hand, sometimes secularists will misuse the language or theory of evolution to make claims about human identity that are false and destructive. For example, a few years ago, a study found that chimps share 99.4 percent of DNA with human beings. One researcher said about this, "We humans appear as only slightly remodeled chimpanzee-like apes."[3] "Only slightly remodeled"—implying, based on the percentage of shared DNA, that there's really not much difference between human beings and chimps. If you really believe that yourself, or if you wonder if that's really true, just ask yourself if you would have a chimpanzee babysit one of your children. Would you elect one to Congress? Would you date one? Would you hold one morally accountable for its behavior? Human identity, the human condition and human worth are huge questions. They're not going to be answered by analyses of the shared percentage of DNA with other creatures. They're not that kind of question.

Several years ago I was at a BioLogos conference. Francis Collins and a number of other scientists who are also people of faith were there. I was struck in talking with them by how lonely so many of them said they were spiritually. I can't tell you how often I'd sit down with somebody at that conference and hear them say, "You know, when I'm at work and I'm with a bunch of scientists, they're really skeptical about my faith. They're suspicious about me." Then they'd say, "When I go to my church, they're really skeptical about me because of my science. I feel like I don't have a place where I really belong." The church ought to be a place where scientists can feel at home.

I want to say to all those who do science, who teach or do research or are otherwise involved in engineering or medicine or education or biology or chemistry or physics or neuroscience: you're doing a noble thing. You are thinking God's thoughts after him. You are reading the big book of creation alongside the little book of Scripture. You are obeying God's command given way back in Genesis to exercise dominion, to learn about, to be curious and discover and steward the earth. Those of us who are not scientists can only shake our heads in admiration. We are grateful and humbled, and we cheer you on. We are so glad you are a part of the body of Christ. We're so glad and grateful and proud to be part of a community with you. Keep on learning and keep teaching us! Be patient with us.

Let's be people who are humbly submitted to the truth. So often in our day there's a misguided notion that some people—secularists or scientists—are open to truth and that faith is about not listening to reason and believing whatever is written down in a book. That is simply not true. The Bible is part of a great thoughtful conversation that went on for a long time with real, thoughtful people. Jesus would be the first person to tell you to follow the truth ruthlessly, wherever it leads.

Has science proven that faith is irrational and God doesn't exist? Not by a long, long shot. Even thinking about, reading and researching these topics gives me such a sense of wonder and awe about a God who could do this. Here's a final quote from C. S. Lewis: "God is no fonder of intellectual slackers than of any other slackers. If you are thinking of becoming a Christian, I warn you, you are embarking on something which is going to take the whole of you, brains and all."[4] We need the whole of us, brains and all, on this journey. God, help every one of us to be open and humble before truth.

12

From Intelligent Design to Evolutionary Creation

Dennis R. Venema

Dennis R. Venema *is professor of biology at Trinity Western University in British Columbia and fellow of biology at BioLogos. His research focuses on the genetics of pattern formation and signaling, and he writes frequently about science and faith. Dennis and his family enjoy many outdoor activities along the Canadian Pacific coast.*

■ ∙ ■ ∙ ■

IT MAY BE SURPRISING TO THOSE who have read my work at BioLogos that until relatively recently I supported the intelligent design movement.

I grew up in British Columbia, where I spent a lot of time in the woods with my father and brother. While my peers at school wanted to be astronauts and firemen, I dreamed of being a scientist some day.

In my local church science wasn't held up as a potential vocation, but it wasn't denigrated as suspect either, and science-faith issues were seldom discussed. Still, I seemed to acquire an anti-evolutionary position by default. I knew of no Christians who accepted it. I even recall dreading hearing the word spoken aloud; evolution, in my mind, was synonymous with *atheism*. Fortunately for me, my high school biology class skipped evolution altogether, but I still found biology to be intensely boring when compared to chemistry or physics. The theoretical underpinnings of biology were missing: a way to organize the laundry list of information into a *context*. Later, I realized that *evolution* was the theoretical underpinning that was missing.

As a high school student I left behind my childhood desire to be a scientist and set my sights on medicine. Biology seemed a natural choice for an aspiring doctor and off I went to the University of British Columbia in 1992.

I soon found that a secular university was not going to crush my faith like I had thought. Rather, I joined InterVarsity Christian Fellowship and enjoyed the friendship of Christian students. Biology, however, remained boring and laundry-list-like. The bright spot was that evolution was hardly mentioned and no compelling evidence was ever discussed: courses seemed more intent on a description of biodiversity than an understanding of how that diversity arose.

As a junior, I became a biology honors student, which meant working on an undergraduate research thesis with a professor and attending an honors seminar class. Experiencing research was electrifying: here at last was genuine science! Not long after, my childhood interest in science blossomed again.

The seminar class included an assignment that required students to familiarize themselves with the research of one of the professors in the department. The work of Dolph Schluter on experimental evolution interested me, and I took the opportunity during my presentation to trot out every anti-evolutionary argument I had ever heard. The class was quite engaged by the presentation, and there was vigorous discussion with students who knew the scientific research better. Dolph had arrived early for his own presentation to the class and heard much of my nonsense, but fortunately he had no interest in embarrassing me. I thought I had scored a victory for the faith against the evils of evolution.

Not long after that, I was introduced to the intelligent design (ID) movement through Michael Behe's essay "Molecular Machines." This essay presented the argument from irreducible complexity that later appeared in *Darwin's Black Box*, a book that I read as a brand-new PhD student. It confirmed in great detail what I already believed: *of course* evolution could not produce anything genuinely novel.

Having read only one ID book and no critical commentary on it, I promptly shelved it and went on to other things. As a graduate student I was extremely busy teaching Mendelian genetics and getting my PhD research going—which meant a large amount of fruit fly genetics. I spent hours glued to my stereomicroscope sorting anesthetized flies with a fine paintbrush. This meant I had a lot of "dead" time on my hands, and so I looked for ways to engage my brain in something constructive while shoving insects around. While many students listened to music, what I found was even better: Regent College recorded most of their classes and put those tapes into the library as items that could be checked out. There for the taking were several decades' worth of classes in exegesis, hermeneutics and the like. Over the next few years, I would eventually exhaust the library collection of Gordon Fee's classes and move on to those of Bruce Waltke, N. T. Wright and many others.

Meanwhile, my understanding of science matured as a graduate student. One frequent activity in grad school was participating in "journal clubs"—groups of scientists and their grad students who got together to dissect papers in their discipline. What one learns in this sort of setting is invaluable. Here I would see papers trashed for their poor experimental design and lack of appropriate controls, or vaunted for their elegance and powerful approach. For the first time, I was approaching science as a (young) *scholar*, not as a student.

After graduating from the University of British Columbia (UBC) in 2003, I took a full-time job as an assistant professor in biology at the largest evangelical university in Canada, Trinity Western (TWU).

This was another busy phase and one of little reflection on evolution—there simply wasn't the time. It was a topic that came up more than it had at UBC: students at TWU were not afraid to ask questions or consider their theological implications. I also had Christian colleagues with settled views on evolution. Most of those in the biology department overtly accepted it, while one colleague in environmental chemistry was an open supporter of the ID movement. Still, the basic approach to evolution I had accepted as an uncritical undergrad and early grad student continued to hold, and I saw no reason to change it.

Later on I was given an opportunity to publish an invited paper. Years before my arrival at TWU, a collection of essays had been written by several TWU faculty members based on the general theme of "A Christian Perspective on . . ." that covered disciplines from art to chemistry. This collection was supposed to be published as a book, but it languished for over a decade. In 2007 a publisher was finally found who called everyone to polish up his or her work for publication. By now the author of the biology chapter had just retired and suggested that I rework the essay as a coauthor. Thinking it was an easy route to an all-important publication, the matter was settled just before I left for the 2007 National Association of Biology Teachers (NABT) meeting.

I was off to give a paper on some innovations I had hit upon for how to teach with fruit flies at a small institution. I had published a paper on this topic the year before, and that was enough to land me a presentation slot, but I didn't know at the time just how relevant this conference would be for the essay I would rework upon my return. The keynote speaker, Francis Collins, spoke on the human genome project and mapping common genetic variation within human populations. Other presenters were connected with the *Kitzmiller vs. Dover Board of Education* trial that had occurred in Pennsylvania two years before. That case had resulted from the Dover Board of Education adopting a pro-ID policy. It tested the constitutionality of teaching ID in public school and was a failure for the ID side. I purchased a documentary about the trial at the conference and watched it that evening.

On the flight home, I realized I knew virtually nothing about evolution or ID. If I was going to write anything credible in this biology essay, I had much work to do. I knew that Behe had just come out with a new book, so I decided to start there. I had heard much anti-ID rhetoric at the conference, and thought it best to look at the case for ID before looking at the case for evolution.

Returning from the conference, I worked on revising the essay. It turned out to be more outdated than I expected. It had avoided the creation/evolution issue almost completely, so there was a lot to be done. In the end, only 10 percent of the original piece remained.

My research began with Behe's new book, *Edge of Evolution* (*EoE*). I wish I had a video of myself reading those opening chapters. It was not long before a frown appeared. I couldn't believe what I was reading. Where was the Behe of *Darwin's Black Box* who had so captivated me years ago? I had fully expected to be amazed watching Behe take evolution down a peg or two, yet even knowing very little about evolution myself, I could already see glaring holes in Behe's argument. Then, when he discussed a topic I was familiar with (population genetics), I knew Behe was out of his area of specialty and depth.

Before I had finished the book I was done with ID. I lost my faith in ID not by comparing it to the science of evolution, but by reading one of its leading proponents and evaluating his work based on its own merits.

Having rejected ID, I began to look into the evidence for evolution. This transition, however, required only ten or fifteen minutes—I read the first research article on my reading list, the 2005 *Nature* paper comparing the human and chimpanzee genomes. The contrast with ID could not have been starker: here was nothing but argument from evidence. As a geneticist, I was fully capable of evaluating that evidence, and it was compelling. My eyes were now opened to the scope of evolution as a foundational theory of biology, and I reveled in the deeper understanding.

I continued to read. As I knew, other Christians had walked this road before, and I found two books especially helpful: *Finding Darwin's God* by Ken Miller, and *The Language of God* by Francis Collins. Though colleagues at TWU counseled me against being "too open" about my new views, I was determined that the essay I was writing would reflect what I thought was the best way to put science and faith together. For better or for worse, I nailed my colors to the mast.

My transition from aligning myself with ID to accepting evolution was rather sudden. Looking back, I realize my training as a geneticist had been invaluable: most evangelicals cannot read the primary scientific literature on evolution. Beyond this, the rich theological material that I listened to as a graduate student helped as well. Through that material I realized that Scripture was interwoven with mystery, tensions and scholarly issues that are not discussed in the average church. Working through some of these issues had slowly but surely washed away tendencies of rigid thinking: I now knew that Scripture had varied genres within it and that the opening chapters of Genesis had the hallmarks of an ancient Near Eastern worldview. As such, the realization that evolution was a well-supported scientific theory did not precipitate a theological crisis for me. My understanding of mystery and tension in Scripture, which many pastors are afraid to touch in a Sunday sermon, was just what I needed to handle this shift. It had created habits of mind that were more at ease with reevaluating long-held assumptions.

An additional factor that eased this transition was that my experience of God had deepened throughout my education. Specifically, my relationship with God was not tied to a certain interpretation of Genesis or a literal mode of biblical interpretation anymore, because I was experiencing his power and presence personally. That experience did not suddenly evaporate the moment I understood the evidence for evolution. Instead, God's empowering presence continued to be part of my life as I explored his creative mechanism, evolution.

Like evolution itself, my path was sometimes slow and at other times rapid, but through it all I have no doubt that this journey was ordained and sustained by my Creator, who has patiently led me to a deeper understanding of his creation. All truth is God's truth, and the book of his works is one that he desires us to take, read and celebrate.

13

A Scientist's Journey to Reflective Christian Faith

Praveen Sethupathy

Praveen Sethupathy *is assistant professor in the Department of Genetics at UNC Chapel Hill, where he directs a research laboratory focused on genomics and human disease. He lives in Hillsborough, North Carolina, with his wife and three children, where he serves as preaching elder for Resurrection Church.*

> *We must know where to doubt, where to feel certain, where to submit.*
>
> **BLAISE PASCAL**

■ ▪ ■ ▪ ■

As Daniel Taylor puts it in his book *The Myth of Certainty*, "being human is a risky business."[1] From the moment we arrive on this earth, we explore the unknown and tackle unexpected challenges. As we are often fond of reminding each other, "with great risk comes great reward." It should not be too surprising that risk is also at the center of our relationship with our Creator. Knowing him is a journey that is necessarily disruptive of our plans, as it intends to take us to new places we could not have imagined. But my experience has been that as Christians, particularly in today's evangelical subculture, our lives often do not reflect this reality. Why are we so alarmingly risk-averse when it comes to our faith? Perhaps since we believe that God is the one sure thing in our lives, we feel uncomfortable with questions that challenge our notions of him (even if there is the chance that they might help to reveal more of him). So we tend to play it safe and build a comfortable nest for our faith, protected from what we perceive to be the threatening cacophony of worldly ideas that rise around us. And to ensure that our faith is not infiltrated, weakened, diluted or lost, we set clear, non-negotiable boundaries for our belief systems, even in areas where the Bible may not demand them. In all this, while we think we are protecting our faith, we may find that we are actually preventing it from growing.

It was during my college years that I became a follower of Christ, in part through campus ministries, which played an important role in my growth as a new Christian. However, these organizations also served as the gateway to a Christian evangelical subculture that, while initially foreign to me, was very compelling because it provided security and a sense of belonging. So without much of the kind of thought and reflection that characterized my conversion journey, I quickly adopted the lens through which the subculture interpreted the world. Perhaps the most striking example of this was with regard to evolution (specifically the idea of common descent), which was commonly viewed within evangelical circles as a notion that threatened the core of the Christian faith. As I transitioned to graduate school to study genetics/genomics, I felt I had a unique opportunity and important responsibility to dismantle the theory of evolution from within the vanguard of academia. But I could not shake the troubling feeling that I was being intellectually dishonest with myself. While I thought I was knowledgeable on the subject of evolution, I was mostly just parroting what little I heard from others, who themselves were certainly not experts on the matter. God opened my eyes to a simple reality: I had not bothered to *reflect* on why I felt this way about evolution, beyond the spoon-fed talking points. Perhaps what I was really defending was my own sense of security within the evangelical subculture. My plans were about to be interrupted by God.

The point of the interruption was not God's setting the record straight for me about evolution. He had more important things in mind. He revealed to me that, despite my intentions, I was not loving him. God wanted me to follow *him*, not any other person or subculture, Christian or otherwise. According to Jesus, this is done in part through the use of my *mind* (Luke 10:27). Jesus freed me to be a reflective Christian—one who believes that God has called us to think and engage, one who sees that faith doesn't require protection but honesty, and one who appreciates that there is more nuance and complexity in life than our need for security sometimes allows. With this new freedom in Christ, I committed myself to study biological evolution and its theological implications. Contrary to expectations that this bit of risk taking would dilute or paralyze my faith, I found that it grew by leaps and bounds.

Today, as a committed Christian called to a vocation in the biological sciences, I have a front row seat to the present-day culture war between science and faith. I have been pitied by some colleagues who view my faith as untested and antiquated, and I have also been perceived as a threat by Christian brothers and sisters who assume that I am advocating science as the ultimate answer. So I am often cornered into the question, "God or science?" from both sides. But I have come to believe that the question reflects an incomplete understanding of the relationship between science and the Bible.

Science provides a set of tools that are useful for probing naturalistic phenomena. As powerful as it may be for dissecting planetary motion or battling cancer, it is not intended for that which is beyond the natural realm. Science is necessarily agnostic with respect to anything outside of the natural realm. It does not accept it, and it cannot refute it. I am not necessarily arguing for "non-overlapping magisteria"; I'm only saying that science cannot support or deny supernatural aspects of our existence.

I also do not believe that the Bible is meant to be a scientific text. In other words, the Bible's primary objective is not to describe the mathematical language, physical laws or chemical makeup of the world. Its goal is entirely different: to speak of God's interwoven presence in the history of humankind, his love for us, our need for him, eternity, sin, redemption and restoration. The Bible communicates these things in diverse ways, through prose, poetry, song, parables, polemics, rhetoric, observational language—in whatever way will help us best understand who God is, what he has done for us and why. Treating the Bible as a scientific text can turn out to be like a robot reading *Romeo and Juliet*—the true meaning and effect may be missed. If we don't humbly and prayerfully strive to understand the richness of the language and the breadth and depth of its intended meaning, then we will be in danger of not only limiting the power of God's Word in our lives, but also of committing injustice in the name of God—such as the church's reaction to Galileo and Copernicus. We have to ask ourselves whether we are making any similar mistakes today—because a high view of Scripture respects both what the Bible is and what it is not.

From ancient rabbinical literature to early church fathers to contemporary Christian scholars, Judeo-Christian communities

have held highly diverse and nuanced understandings of Genesis
1–2. In my own study, my goal has been to treat the text with the
respect it deserves, which means appreciating its style of language and original cultural context, rather than reading into the
text my own modern biases and preconceived notions of what
it means to say.

So where does this leave me on the topic of evolution and
creation? I describe myself as an *evolutionary creationist*. If this
seems like an odd juxtaposition of terms, I believe that is largely
because the voices that speak against the possibility of harmony
between science and faith are often the loudest in our culture.
These voices have popularized the notion that evolution is inher-
ently atheistic, an assertion that conflates biological evolution
with ideological evolution (naturalism). Biological evolution is
technically agnostic with respect to God. Indeed, Darwin's thesis
was titled *The Origin of Species*, not *The Origin of Life*. Many who
see the evidence for biological evolution choose to believe that
God is not the source of life. But that is their *personal belief*—the
evolutionary process itself does not compel them either way. To
me, evolutionary creationism means that I affirm God as the
Creator of the world, and that his mode of creation involved biological evolution.

One challenge I faced is the perception that common descent
threatens the Christian belief that humans are uniquely made in
the image of God. I have come to share the belief of many Christian
theologians and Bible scholars that we are not compelled to understand "image of God" as physical or material. Rather, I believe
that it is our God-given spirit, desire for fellowship with God and
God-ordained royal priesthood (1 Peter 2:9) that differentiates us
dramatically from all other created beings. Evolution describes
biological change, not our *spiritual* nature. Therefore, it appears
to me perfectly harmonious to believe that after bringing about

the biology of *Homo sapiens* through the evolutionary process, God then set them apart by giving them spiritual sensitivity and purpose (that is, he made them "in his image").

In the spirit of intellectual honesty, it is worth mentioning that I have not worked out every jot and tittle of every possible theological implication of common descent. I am still very much on the journey and expect to be for my whole life. My primary passion is not to teach evolutionary creationism to Christians; rather, it is to reignite within our Christian communities a passion for thoughtful reflection, honest inquiry and the pursuit of knowledge. Yes, this might be risky, because thinking can be threatening to the comfortable lines we have drawn around our faith. But if they aren't God's lines, they're just holding us back. So let's ask questions together, thereby allowing one person to sharpen another (Proverbs 27:17).

14

A Fumbling Journey

Dorothy Boorse

Dorothy Boorse *is professor of biology at Gordon College and lead author of* Loving the Least of These: Addressing a Changing Environment, *published by the National Association of Evangelicals. She has an MS in entomology from Cornell University and a PhD in oceanography and limnology from the University of Wisconsin-Madison. She loves wetlands and lives with her family in Beverly, Massachusetts.*

■ · ■ · ■

WHEN THE CLASS ENDED, A STUDENT was crying right in the front row. Two or three people hovered around her, one handing her a tissue, another giving a hug. We had been discussing the fact that the earth looks really, really old and that many Christians believe this is consistent with the purpose of Genesis and its ancient Near Eastern literary style. "But I don't know what to think," she said. "My professor who told me otherwise was so godly." Her distress was profound, and it brought back to me the journey I have been on as well—a journey in which I have discovered that for his own good reasons, God allows even godly people to be wrong about points of fact.

I remembered back to my own school years when a shy, kind man with thinning hair taught middle school science in our small Christian school. A poster hung on the wall. It showed a progression of supposed prehuman forms. Under each was an explanation of why it was a fraud. The poster mocked the entire idea of evolution— using the Piltdown Man as Exhibit A, but also calling many other scientific findings frauds. The poster suggested that the scientific fraud was so obvious that a twelve-year-old like myself could see it, but so difficult to understand that the entire scientific community was befuddled. My friends and I loved the idea, tickled that we somehow knew something that the experts didn't.

Misinformation, however, was not limited to the poster. I clearly remember a slide show, projected on the wall from an ancient projector. The slide series purported to show proof of a young earth which most of the believers I knew supposed to be described in the Bible. In the slides were pictures of what I was told were dinosaur tracks alongside human tracks, seen in Glen Rose, Texas. "Wow," I thought. "Look at that! We know something that they've missed." The "we" referred to Christians, and the scientists were "they." On that day, with a fly lazily making its way through the class and the poster curling on the wall, the hum of the projector and the slides of rocks along the Paluxy River showing on the wall, I did not suspect how much my opinion would change. Within only a few years, the Glenn Rose footprints were revealed to be a hoax: some had been deliberately carved out of rock, others misidentified. A movie made about them would later be retracted. The very idea of human and dinosaur footprints in Texas had been debunked and I myself was faced with overwhelming evidence that my elementary and middle school science had been flawed.

My family was a learning one—we learned the names of plants, dissected the occasional dead mole or rabbit, discussed the role of tannins in the ecosystem of the Pine Barrens and read broadly. My father, a biology teacher with seminary training, was not as easily impressed by materials we brought from school. "Some biblical scholars believe the Genesis account does not actually mean that the days were twenty-four hours," he told us. "Many Christians have no problem with science and Christianity." We were given space to believe evidence that the earth was old. A good family friend was a geologist and a Christian. At home I pondered how these things might fit together.

Somehow, by high school I found my position shifting. I remember one scene in particular. My family was visiting a different church one Sunday. My older brother and I were in the high school Sunday school class. Evolution came up, and the students and Sunday school leader were saying things about evolutionary theory that I knew did not make sense. Without intending to, my brother and I found ourselves disagreeing with the group. If you are going to contradict something based on science, we both felt, you need to have logical arguments. The discussion got awkward and I wondered how I had ended up there. I was very devout. Why was I the one sitting in this church basement youth group room with people I did not even know, sticking up for a part of science I still knew little about? To them it sounded like I was disputing Christianity. How had that happened?

In college, many things came together. I was exposed to a number of different ways to view both Scripture and current science and was able to consider the different views for myself. I could see ways science and faith might relate without conflict. My relief was palpable. It was like a tight spring uncoiling. I got to know many people who were able to put things together that I had seen in tension. But it wasn't always easy; I had my own experiences, just like my student would have years later in class. I discovered that I sometimes disagreed with people whom I loved and respected. And it wasn't just about science. I found out that a kind family friend couldn't believe women were capable of leadership, although he personally cared deeply for my sister and me. I remember an elderly woman in my church explaining that she gave up cards and dancing when she became a Christian, while I privately thought both were benign activities. I saw people who were kind, loving, complicated and sometimes, just sometimes (at least in my opinion), wrong. In these experiences, I was able to see that you could disagree with people and still love them. This epiphany, that loving people is separate from agreeing with them, had been modeled for me by a number of my mentors. I especially saw it in faculty at the Mennonite high school I attended and in my parents. Indeed, my middle school teacher himself also modeled this very trait in many subjects—but it was only in my young adulthood that I began to understand the idea.

Even later in college, however, I had a crisis of faith. It was not primarily precipitated by evolution or other issues of science and faith, although they were present. I struggled, as many do, with the problem of pain and suffering in the world, the environmental degradation I saw around me and the terrible history of human actions toward each other. I could not understand the history of Christianity or why people of faith could do so many unjust things. I saw my own sins and weaknesses. I felt that we were trapped. The human condition was terrible, and I doubted whether God could be both powerful and good. It is hard to explain all of the ways I needed to heal, but I did. As I came through this period, in which I struggled with depression as well, I made a conscious choice to live in joy. Because I found the words and life of Jesus so compelling and because I had such a strong sense of my own need for forgiveness, I decided again to follow Christ. I have continued in my faith even as I pursued higher education in the sciences.

I have the privilege of studying ecology, a discipline in which competition, symbiosis, natural selection and adaptation are central. So is the concept of limits. God made a world in which materials and energy are limited, but God himself is not. The natural laws of the living world drive species to differentiate and to fit their changing environments. There are many disputes about evolution. Aside from broadly saying that God used a lot of evolution in the creation of species, I won't remark on any details. Some I don't even have an opinion on. However, it is helpful for me to look at evolution the way I see laws of physics, such as those driving evaporation, plate tectonics or the force of gravity. Evolution is similar to water poured on the ground of a hillside: as it rolls downhill, it forms rivulets. We may not be able to predict the exact placement of any one tiny steam as it passes a clod of soil, but we know the water will subdivide and we know that it will run downhill. Take another example: evolution is unpredictable in the same way that, although the actual positions of gas bubbles in a boiling pot may be unpredictable, the temperature at which it will boil is clear. There are laws that govern living things, and the laws of survival in a changing environment are one such set.

I now have an internal sense of wholeness that comes from feeling that the science I understand fits with my dearly held beliefs. Unfortunately, however, as an adult I have found that my world is often divided. In churches I hear one thing, in the secular world another. Dear friends dismiss what I know with a wave of their hand; scientists on list serves mouth off about people of faith.

The rancor exhibited in public debates on almost any topic is painful. In order to live in this polarized world and to have a complex, nuanced view, I've taken some approaches that I think are imperative. The first is that I agree with people when I agree with them and disagree with them on issues where I don't. That is, I won't disagree with someone about one issue simply because I disagree with them about something else. People are complex, much more so than our often-bundled positions allow.

I also made a decision to love everyone regardless of how much we agree. I believe strongly that this is biblical. In Paul's lyrical description of love in 1 Corinthians 13, we are told,

> If I speak in the tongues of men or of angels, but do not have love, I am only a resounding gong or a clanging cymbal. If I have the gift of prophecy and can fathom all mysteries and all knowledge, and if I have a faith that can move mountains, but do not have love, I am nothing. If I give all I possess to the poor and give over my body to hardship that I may boast, but do not have love, I gain nothing. (1 Corinthians 13:1-3)

Notice the gifts we are told about: speaking, prophecy, knowledge, faith and generosity. These are vital qualities. However, while all of these matter, none matters outside of love. Even being right is not the end goal. Knowledge and being able to fathom mysteries have no meaning apart from love. Our goal is not simply to be correct, but to discover truth and express it in love.

This love isn't always easy. For example, as a teacher, I meet some students and their parents who think, at least initially, that a central concept of my whole discipline is founded on an error. While part of my job is to teach correct science, I must also strive to relate to them as fellow Christians, praying with and encouraging them. I encounter similar situations at church or in volunteer work. I often fail in my ideal of loving first, but I am committed to that as the goal.

While evolution is an important concept in biology, it is not what I spend the bulk of my time on. In my current work I try to educate people about care of creation, and the terrible impact of environmental degradation on the poor and on God's other creatures, over whom we watch. The reality of environmental problems can also cause difficult discussions. Many times the public conversation is testy and people do not give or receive the benefit of the doubt. In this context, it is helpful to remember the importance of truly listening, and of seeing the people you communicate with as recipients of God's love first, and as people you want to persuade second.

Years after I began my fumbling journey, trying to understand the book of the natural world and the book called the Bible, I've reached a point of dynamic peace. That is, my understanding of how science and faith fit together changes, but I believe that they do. All truth is God's truth. As my tearful student discovered, people we love and respect can be wrong. And the reverse is also true—we can love and respect the people we believe are wrong. So let's work together on the project we have been given—to better understand the natural world and better understand the Scriptures—two ways of honoring God.

15

A Biblically Fulfilled Evolutionary Creationist

J. B. (Jim) Stump

J. B. (Jim) Stump *is senior editor at BioLogos. His most recent book is* Science and Christianity: An Introduction to the Issues. *He and his wife have three sons and two cats. They enjoy books, sports, music and traveling.*

■ ∙ ■ ∙ ■

IN THE FALL OF 2009 I WAS AT the Evangelical Theological Society meeting in New Orleans. Browsing the book room just before heading home, I picked up a copy of the newly released *The Lost World of Genesis One* and put it in my carry-on. I'd never heard of its author, John Walton, but as a Christian philosopher with an interest in issues of science and the Bible, I thought it might be a good read. I don't think I set the book down on my flight back to Chicago. I was mesmerized and could sense that some order was being imposed on my mind where there had been pockets of disorder.

My father started his career as a middle school science teacher, and although we were conservative Christians I never felt threatened by scientific findings the way many in my community did. My siblings and I were taught to value the natural world. We were never discouraged from asking questions. Our family had a sense that Christians should be identifiably different from the world, but not by withdrawing from it. We were in the world, not of it.

In my public high school, I don't remember one class period where evolution was discussed, and the science classes at my Christian college avoided the topic too—though the apologetics classes made a pretty big deal about it. I took an undergraduate degree called science education, but focused primarily on mathematics. I liked the logical and coherent systems of mathematics, and there were no tensions for me between math and my Christian faith.

After college, my wife and I taught at a missionary kids' school in Sierra Leone. This was before there were Internet cafés or cell phone service in remote African villages, so after the sun went down there wasn't much to do except read by lantern light. I made it a good way through the school library's shelf of nineteenth-century literature—Tolstoy, Melville, James and (my favorite) Dostoyevsky. Somehow when the analytic tools of mathematics met the grand ideas of this literature, a desire to study philosophy popped out.

In graduate school I was drawn to philosophy of science classes, which helped me separate the science of evolution from the ideology that is too often bundled with it. Once I was exposed to the evidence I didn't really have a problem affirming evolution, and I don't have an interesting "conversion" story in that regard. Still, as a Christian who recognized the authority and inspiration of Scripture, I wasn't always sure how to reconcile science and the Bible. I wasn't persuaded by those who claim, "The Bible obviously says . . ." as an argument against evolution. To my mind, these people have to do a fair amount of gerrymandering to get the biblical witness to say such things, and their "obviously says" hermeneutical approach has to be completely abandoned for other parts of Scripture (such as when the Bible obviously says we should not mix fabrics in our clothes, or that we should greet one another other with a holy kiss). But I didn't really know what to say besides, "The Bible must not mean that, because the natural world clearly testifies otherwise."

Richard Dawkins has gotten a lot of mileage out of his claim that Darwin made it possible to be an intellectually fulfilled atheist. I'd like to redeem that claim and transform it into something better. Reading Walton's book helped me become a biblically fulfilled evolutionary creationist. In *The Lost World of Genesis One* I found a way to think about the Bible from our cultural standpoint today. It is clear to me now that we must consider the ancient Near Eastern world when we interpret the text of the Old Testament, and I'm persuaded that if we're going to take the Bible seriously, there is a lot of study required to understand its broader cultural context.

I have spoken on the theme of science and the Bible to a number of churches and other Christian groups. It is almost always helpful for such audiences to see concrete examples of misunderstandings of Scripture that can be corrected by a better understanding of the ancient context. On this point I'm basically a Walton impersonator. One of Walton's examples concerns the fairly recent discovery of ancient ziggurats in the Middle East (massive pyramidal structures that were part of temple complexes), which has led to a better understanding of the Tower of Babel story in Genesis 11. I used to have the flannelgraph version of the story in my mind. According to that version, these people were building the tower so they could climb up into heaven. I was always a little confused by this, because it didn't make sense to me that God would be worried about that. Was this somehow an achievable goal? But the cultural context of ziggurats gives us a much better interpretation of the story. The tower was intended to bring God to earth in that location—that was the point of ziggurats in the ancient Near East. God's frustration of their efforts, then, is to distinguish himself from the cultural expectations: "This is not how I will come to dwell among you. I am not that kind of God."

There are many other such examples of cultural context illuminating the meaning of Scripture. Having this richer understanding helps clear the way for more constructive discussions about the Bible and evolution. It shows the error of taking the "plain meaning" as the definitive interpretation, and it cautions against concordist approaches to Scripture, according to which we can read modern scientific concepts into the text. For example, some people want to turn Genesis 1:20 and 1:24, where God says, "Let the waters bring forth swarms of living creatures" and "Let the earth bring forth living creatures" (NRSV), into endorsements of the science of evolution. We might take the "bring forth" clauses as evidence that the ancient Near Eastern mind had no problem with God creating through natural processes, but it is a concordist error to see these passages as unambiguous descriptions of evolution—or other passages as descriptions of the Big Bang or even heliocentrism. Those concepts weren't available to the ancient Near Eastern mind, and God wasn't concerned to reveal modern scientific concepts to the ancient human authors of Scripture. God revealed himself and his will to the human authors in terms they could understand. So again, in order to properly understand the text we need to know about the ancient thought world from which the text emerged.

When I make this point there is usually someone in the audience who expresses despair, afraid that they need a PhD in order to read their Bible. Thus I think it is important to make two further points about interpreting Scripture: First, every person can benefit from picking up the Bible and reading it every day. We evangelicals believe that God speaks to us through the Bible—he doesn't just speak to people with PhDs. The Word of God is living and active for all of us, but it is easy to hear God's voice only in the key to which our ears have been tuned. So, second, every community of believers needs Bible scholars they trust to help them read the Bible better. At least since the Protestant Reformation there has been a tendency for Christians to develop their own individual interpretations of the Bible—and form their own denominations! Scripture is a gift to the church, and the church is a community. In that community we must recognize that some people are better trained and equipped to interpret Scripture, just as we recognize there are experts whose opinions should be respected in the fields of medicine and auto repair. So our individual readings should have regular guidance from experts.

I'm aware that there is disagreement among Old Testament scholars on some of the details of Walton's position on Genesis 1. And there are, of course, lots of other issues of biblical interpretation on which the experts disagree. My one semester of Hebrew does not qualify me to speak authoritatively on the specifics of these disagreements. But I'm not claiming that Walton's position must be correct in every aspect in order to reconcile science and the Bible. I'm merely saying that his work has been a gateway for me (and many others) to consider a more sophisticated treatment of Scripture. This sort of work has also given me confidence that we can take both Scripture and the natural world seriously.

That confidence gave me permission to really dig into the academic study of science and Christianity. On a confidential survey question I would have answered that I accepted evolution and that I was a Christian, but I had not really wrestled with the implications of those beliefs. I suppose I was nervous about what I might find. Reading Walton's book and discovering BioLogos helped me put away those fears and embrace the model of evolutionary creation. But as the chaos in my mind was brought under control, the implications of openly endorsing evolution brought about a different kind of chaos.

One of the arguments often used by anti-evolution crusaders is that we Christian academics who accept evolution do so out of a desire for professional recognition and advancement. Such a claim is curiously at odds with any facts I know. Christians who have positions at secular institutions have nothing to gain by openly embracing evolutionary creation. In fact, it may very well count against their professional standing if their religious commitments become known. Professors at Christian institutions risk professional suicide if they come out as evolutionary creationists—as I know from personal experience.

I taught at a conservative Christian college for seventeen years and was very much an insider there. Two of the buildings on campus are named after my ancestors, and my children are third-generation students of the institution. I was widely viewed as a popular and successful professor, both on campus and by many influential constituents of the institution. Faculty members were only ever required to sign a fairly generic statement of faith along the lines of the Apostles' Creed (for example, "God is the creator and sustainer of all things"), and there were professors both in the sciences and in religion and philosophy who accepted that evolution is the best scientific description we have for how God created the world. I know this because we talked about it—usually in hushed voices behind closed doors.

After my newfound confidence in reconciling science and the Bible, I started speaking more openly about the topic of evolution. I never attempted to indoctrinate students, but presented the evidence for how Bible-believing Christians could accept the science of evolution. At one point I used the topic of creation and evolution for a critical thinking class (after receiving the administration's approval) in which I presented the strengths and weaknesses of a range of positions and guided students in thinking carefully about these positions (in these classes I never revealed which position I myself held to—it was a critical thinking class after all). Because there were a few parents who raised some concerns about this, the administration thought it prudent to begin a broader dialogue about origins with faculty and the leaders of the affiliated denomination.

Over the next two and half years I was involved in many of these discussions, and it became vividly apparent that there was a wide gulf between those of us who had studied origins academically and the church leaders who spent most of their time on areas of practical ministry within the church. I'm convinced that both sides were motivated by the goal of serving God and his church faithfully, but there were very different understandings of what that looked like in our respective contexts. At the college we believed that a commitment to Christ and to the essentials of the faith could provide safe spaces where students and faculty could openly pursue difficult questions and thereby push faith to deeper and more mature levels. The church leaders believed these kinds of questions undermined the authority of Scripture.

This is a real public relations challenge for Christian colleges. So many conservative Christian parents have been persuaded to make a college's stance on origins a major criterion for where they'll send their kids to school. In the end, even though I had a long and positive standing in the college and denomination, the tide turned against me and the college's board adopted the denomination's narrow origins statement. I could have stayed if I'd agreed not to publicly contradict this new statement, but I felt it was best to resign my position and invest my talents for the kingdom elsewhere.

My story is not unique. There has been and will continue to be tension between science and Christianity, especially on Christian college campuses. I don't think it is proper to explain this tension as the educated and enlightened professorate versus the ignorant masses. Belief structures are complicated, interrelated webs, and you can't just replace one isolated section of the web without affecting other parts. Because the rhetoric from anti-evolution voices in our culture has done a masterful job of bundling evolution together with all manner of evils, we cannot convince people of evolutionary creation by merely laying out the scientific evidence. It takes time to sort out the implications for one's belief system. But my story is evidence that we can sort these implications out by attending carefully to the Bible and to the natural world. I cannot help but believe that evolutionary creation is a faithful, evangelical option for understanding origins.

16

A True Read on Reality

Daniel M. Harrell

Daniel M. Harrell *is senior minister of Colonial Church in Edina, Minnesota, and the author of* Nature's Witness: How Evolution Can Inspire Faith. *He holds a PhD in developmental psychology from Boston College and has been active in the science and faith dialogue for many years. A Southerner by birth and a New Englander by choice, he is now a Minnesotan by the will of God. He lives with his wife and daughter in Minneapolis.*

■ ▪ ■ ▪ ■

GROWING UP IN THE SOUTH, evolution wasn't something you brought up in polite conversation. And Lord knows you never talked about it in church. It's not that my church necessarily rejected evolution; it was one of those liberal mainline Southern churches where the deacons smoked outside between services. It was more that admitting you descended from monkeys struck too close to home.

My church wasn't exactly "Bible-believing" either. We celebrated Christmas and Easter of course, but Santa was the star in our annual Christmas pageant and the big event of Holy Week was Saturday's egg hunt. Still, I always liked reading my Bible, a

King James Version my dear Sunday school teacher Bertie gave me in second grade. She told me that these words had life in them, and that I should make sure and chew on them every day. By the time I made it to college, my daily Bible reading had connected me to a network of serious-minded Christians for whom faith and life were intricately entwined. They were an ethically minded bunch too, noticeably demarcated from your average college student by their sobriety and virginity. I admired their passion for the poor, their concern for the earth, their zeal for authenticity and their simple love of life. As I fell into their faithful circle, I found that their ethics and passions were fueled by their own reading of the Bible, a reading that exalted God as Creator and thereby rejected evolution. The world described by science—old, random, merciless and meaningless—could not be the world made by the Bible's God.

Working around the scientific evidence was not so easy to do. I tried by taking an astronomy class for my science requirement, assuming that would save me from the indictments of fossils and DNA, but the stars couldn't lie. The universe was billions of years old. To alleviate my growing cognitive dissonance, a campus pastor taught me the trick of "apparent age." He told me that God created the world already mature, just like he made Adam as a full-grown man. The universe looked older than it really was on purpose. It sounded awfully tricky, but who was I to doubt the mysterious ways of the Lord? The trick lessened my dissonance, and as a psychology major, I appreciated the importance of reducing cognitive dissonance.

Remarkably, I was able to trick myself all the way through seminary, years of pastoral ministry and a PhD program in psychology. By then, psychology had changed from my undergraduate years. It was less about observing and controlling cognitive and behavioral *outcomes*, and more about neurons and brain function; that is, observing and controlling the causes of cognition and behavior. Psychology now emphasized the physiology of thought and behavior as its prime locus for study, having adopted an organic, biological model over an information-processing model. A biological model of the mind links psychology to human development, which is inextricably linked to human evolution. Suddenly the question before me, both as a pastor and a psychologist, was what we meant by "mind" or "soul" if those notions were now understood mostly as material brain functions.

While the topic of evolution still wasn't deemed appropriate for church, I started to worry that the college students in the university congregation I served struggled with the same dissonance. The "apparent age" trick was no longer adequate. Not only did it fail scientifically, but it failed theologically too—it seemed to portray God as an intentional deceiver. This would not do.

Integrating science and Christian faith proved a struggle. I tried to preach about how entropy was a consequence of the fall, but physicists in my congregation were quick to correct me on that. Had there been no entropy in the Garden, Adam and Eve would have been up to their necks in bacteria and bugs. I also tried to draw a distinction between a person and her clone using the distinction drawn in the Nicene Creed between "begotten" and "made." Analogous to the way the human Jesus differs from other humans, might not a person born naturally differ from a cloned person? But biologists set me straight there as well. A human clone is genetically the same as one's identical twin.

This latter misstep wasn't made in front of a forgiving congregation, but before a skeptical crowd at an MIT science conference. I was invited to serve as a "religious voice" at a conference otherwise populated by accomplished faculty and clinicians, pharmaceutical executives and congressmen. Human cloning was a particularly hot topic at the time, and I was assigned to a panel to discuss cloning ethics. The auditorium was packed. Spotlights brightly burned the stage where four chairs were parked behind microphones. In front of the seats were name cards: *Nobel Prize Winner for Physics*; *Faculty Chair, Brown University Biology Department*; *Bioethics Professor, University of Pennsylvania*; and then me, *Insignificant Church Minister.* The moderator welcomed the audience and invited each panelist to give opening remarks. What were our positions on human cloning?

Cloned animals are a regular staple of human diets these days, and some countries allow the cloning of human embryos for stem cell research in medicine. Screening of in vitro fertilized embryos for a limited number of diseases and abnormal development already occurs prior to implantation, and human eggs and sperm are available online from Ivy League donors (presumably making it easier for babies to get into Harvard some day). But there remains a moral consensus against birthing fully developed human clones for both developmental and psychological reasons. There's also what ethicists regard as the "yuck factor." Cloning yourself is gross.

Nevertheless, the bioethicist on our panel challenged the idea that cloning is unethical just because it doesn't feel right. What's wrong with duplicating genes? Your body does it naturally with millions of new cells everyday—and that's just to keep your skin healthy. He said that eventually people will get over the yuckiness, just as they got over it with in vitro fertilization. Then "replacement babies" will be as normal as "test tube babies" are now.

It was up to me to offer the "religious" response, which I tried to do with my Nicene Creed analogy. I suggested that children "begotten" as a gift of marital love are distinguishable from children "made" as cloned products of personal preference. A cloned child could not be a "replacement baby" if that meant the same baby, recreated. A clone is only a genetic match, not a duplicate person. A clone could not be the same as naturally reproduced offspring either, because a mother's produced clone is her identical twin sister. Yuck.

A young man stepped up to the microphone in the aisle and asked, "Would a clone nevertheless have a soul?"

Here's where I messed it up. I was prepared to argue that a clone wasn't *offspring*, but I was not prepared to argue that a clone wasn't *human*. "It's a mystery," I replied elusively, sounding very much like a theologian.

The bioethicist seated next to me was visibly agitated. Younger than me, he was probably bucking for tenure, looking to make a splash by taking down a minister. "There's no mystery to any of this!" he bellowed. "Humans are humans are humans regardless of how they come about. Nervous systems are just like respiratory or digestive systems. If something as ridiculous as a soul actually existed, then sure, a clone would have a soul since whatever we mean by soul would have to be the same in every person, and a clone would be a person. The very idea of a soul is nothing more than a fabricated belief conjured up by our ancestors for the sake of some reproductive advantage."

"Hmm, now that is mysterious," remarked the Nobel laureate on my left, who came to my rescue. "Whatever is meant by 'the soul,' it is not solely a neurological entity. Natural selection requires interaction with environments. Cultures and communities play a role too. It's complicated."

And it remains complicated. Following that experience, I read everything I could about evolution and Christian belief, and even ended up writing a book about it myself (*Nature's Witness: How Evolution Can Inspire Faith*). That got me invited into a number of conversations, including one in the rural Midwest where I encountered Christians convinced that science had it all wrong. If the Bible is wrong in one area (age of the earth) it would have to be wrong in every area (the resurrection of Jesus). The breakdown in this logic is its failure to account for the shortcomings of human understanding. Belief in an infallible Bible does not make us infallible people. The same applies to science. While scientific observations of data may prove indisputable, scientific *interpretations* of data get disputed all the time. The same is true of philosophical and theological interpretations. My concern is that differences of opinion between Christian faith and science be understood as differences of interpretation rather than devolving into arguments over the age of a rock or whether a particular gene can be traced back to ancient, prehuman hominids.

It's unfair that so much vociferous resistance to Christian faith has been fueled by a certain interpretation of evolutionary evidence. Scientists themselves have been known to bristle at antagonistic atheists' claims against religious faith that use evolution for support. Evolutionary theory has no stake in the existence of God. Miracles and resurrections and souls and the Bible's authority all lay outside the realm of scientific investigation.

The author of Hebrews insists that "faith is confidence in what we hope for and assurance about what we do not see" (Hebrews 11:1), but that only means that there's more to reality than we see. It doesn't mean we should ignore what we *can* see. Faith is not fantasy. It corresponds with the way things really are. Inasmuch as "all truth is God's truth," any pursuit of truth, through whatever discipline we pursue it, will eventually lead to God. Rather than feeling threatened and frightened by scientific advances, we should see scientific advancement as new vistas for theological consideration—new mountains to explore.

Sure, I didn't like the bioethicist's interpretations or attitude, but that didn't mean he didn't have his facts straight. If all truth is God's truth, a true read on reality will only buttress theological understanding. Clinging to false notions about how God operates in nature only forfeits the opportunity to praise God for how he truly operates (a point Mark Noll nails in *The Scandal of the Evangelical Mind*). Theology needs to function alongside scientific reality or it ends up being not only irrelevant but boring too.

As a minister I don't want to preach about a God who's unrelated to observable reality. If God has nothing to do with life as we live it, then ethics function solely on the basis of utility instead of principle. If God has nothing to do with morality, then principles are nothing but self-generated and self-serving preferences. If God has nothing to do with evolution, then its valueless assertions are free to justify all sorts of aberrant behavior. Evolution was cited as justification for Nazi eugenics. More recently, some evolutionary psychologists have suggested that rape may be a natural behavior (see Frank Ryan's book *Darwin's Blind Spot: Evolution Beyond Natural Selection*). Without systems of faith and value that address actual behavior and choices, it's hard to argue against this.

I believe that "the earth is the Lord's, and everything in it" (Psalm 24:1). Theology shouldn't merely withstand scientific discovery—it should celebrate it as a display of God's handiwork. And not just celebrate, but safeguard it too. Science is too easily tempted by its own sense of importance to abuse its discoveries for power and profit. Science needs the values that theology provides to funnel its work into soliciting wonder at the marvels of creation and into serving the needs of humanity.

17

A British Reflection on the Evolution Controversy in America

N. T. Wright

N. T. ("Tom") Wright *is an Oxford graduate in classics and theology and the author of over seventy books. He taught New Testament for twenty years and served the Church of England for sixteen before moving to St. Andrews University in Scotland as professor of New Testament. He lists music, poetry and golf among his recreations.*

■ ▪ ■ ▪ ■

WHEN I WORKED AT WESTMINSTER ABBEY, one of the questions most frequently asked by visitors, especially Americans, was "Is it true that Charles Darwin is buried here?" On one occasion, noting the route the visitor in question had just taken to walk through the abbey after evensong, I replied, "Madam, I think you just stepped on him." "Good!" came the emphatic reply, which told me something about the visitor in question. However, on another occasion, walking past Darwin's tomb, I

Excerpted from "Healing the Divide Between Science and Religion," in N. T. Wright, *Surprised by Scripture* (New York: HarperOne, 2014). Used by permission.

spotted a little pile of flowers and greeting cards. They were obviously from schoolchildren, and the general tenor of their message was "Mr. Darwin, we love you."

I have often wondered what they had been taught. Can it really be that teachers tell the story of Western culture in terms of pre-Darwinian gloom, superstition, prejudice and the dead hand of religion, with Darwin personally ushering in a new era of happiness, liberation, knowledge and the milk of human kindness? If that isn't a highly selective and oversimplified history, I don't know what is. What's more, Westminster Abbey attracts thousands of visitors from every corner of the world; how come it seems to be mostly Americans who are interested in him and who instantly take sides in an assumed war in which his very name is a battle cry?

I offer this short disclaimer. I realize that I am British (rather than American) and a theologian (not a scientist), and that I am therefore an outsider to this discussion. But I hope that I might point out three things in particular that an outsider may perhaps see more clearly than an insider.

First, I want to point out that the way the science and religion debate is conducted and perceived in North America is significantly different from the ways analogous debates are conducted and perceived elsewhere. Second, I want to suggest that this is at least partly because of the essentially and explicitly Epicurean underpinnings of the social self-understanding of the United States since the late eighteenth century—and that the standoff between science and religion in America is therefore analogous to, and indeed bound up at quite a deep level with, the standoff between church and state, or religion and politics, or however you like to put it, so that you can't address one of these topics without implicitly addressing all of them. Third, I want to propose that we therefore need a much more radical rethink of the underlying worldviews we are dealing with than we have normally contemplated in our science and religion discussions. That, I hope, is the point at which the deeper contribution of a biblical theologian might be useful.

We in the United Kingdom never had a Scopes trial. We did, granted, have the notorious public debate at Oxford in June 1860, between Samuel Wilberforce, then bishop of Oxford, and the scientist T. H. Huxley. Within a generation, the story of this debate had grown and been shaped by a tradition so strong that that tradition has come to be accepted as true, though more recent research indicates that matters were by no means as clear cut as the received narrative would suggest.

According to the tradition, Wilberforce at one point asked Huxley whether he claimed descent from apes on his grand-father's or grandmother's side, and Huxley retorted, more or less, that he would rather be descended from an ape than from someone who so abused his intellectual gifts. This, however, is the stuff of legend. The philosopher John Lucas pointed out some while ago, and this has been taken up by Stephen Jay Gould, that the account in which the agnostic Huxley struck the great blow for freedom from ecclesial dogma took root and spread at a time when the English middle classes, still anxious to gain political standing to rival the aristocracy, had a particular interest in the view that one's pedigree was irrelevant to one's moral worth.

In addition, by the end of the century the world of science had changed significantly, from a sphere in which anyone and everyone might take part (provided they could afford it) to a much more professionalized guild. The picture of Free Science triumphing over the stuck-in-the-mud church fitted the increasingly independent mood of scientists in the 1890s, when some of the key texts about the Wilberforce/Huxley incident were penned, much better than the more freewheeling 1860s. But there, in the late Victorian era, the matter rested, and most people in today's Britain simply have a vague idea that the church is inclined to obscurantism, and that science has set us free from the shackles of its dogma and ethics. The horror of two world wars, with the Great Depression in between, gave people much more to worry about, and indeed far stronger reasons to question the roots of their traditional faith. So, to repeat my earlier point, my sense today is that few people in Britain abandon their faith because of what science may say, though some who left it for other reasons or never had it in the first place naturally find it convenient to retell the stories not only of Wilberforce and Huxley but, farther back, of Copernicus and Galileo and the rest.

In America, however, the Scopes trial clearly had a massive impact, which resonated much more widely into the culture and accelerated a polarization that has not affected the rest of the world in the same way. I recently reread a devotional classic that I had much enjoyed in my early teens: Isobel Kuhn's *By Searching*. When Kuhn was a student in the 1930s, her science professors were scathing about anyone who believed not only in a literal six-day creation, but in any of the Bible, including Jesus and Christian origins. The pressure from professors and peers to capitulate to the pushover science-therefore-atheism position was intense. And I suspect that, as usual, there was far more going on than simply a straightforward, rational—or even rationalist—public discussion. The modernist movement was in full swing culturally and politically, and it was assumed that Christianity, not only but not least its creation account and its belief in miracles, especially those of Jesus, was part of the premodern world that the forward move of history was leaving behind. In particular—and this, I stress, is an outsider's perception—it seems to me that the cultural polarization in American society, including the fundamentalist-modernist controversy of the first half of the twentieth century, has roots that go back at least as far as the Civil War in the 1860s.

Only that kind of backstory can explain the enormous interest generated across the United States by the Scopes trial of 1925, with a three-time presidential candidate (William Jennings Bryan) speaking for the prosecution and reporters flocking to Dayton, Tennessee, to cover the show. It was the first trial in the United States to be broadcast on national radio. But the Scopes trial cannot have generated the standoff between science and religion in and of itself; it merely brought it to a sharp, polarizing moment. And the trouble with sharp, polarizing moments is that they become iconic. Like martyrdoms ancient and modern, and indeed like civil wars, they generated loyalties and counter-loyalties: you *must* now take such-and-such a line, because otherwise you're a traitor.

The words of Clarence Darrow, leading the team in Scopes's defense, have a contemporary ring for anyone who reads Richard Dawkins and his ilk: "We have the purpose of preventing bigots and ignoramuses from controlling the education of the United States." The vitriol of a leading journalist such as H. L. Mencken can hardly be accounted for except on the assumption that the trial was taking place at a major cultural fault line, with supposedly sophisticated city types from the Eastern Seaboard looking with disgust and horror at the "morons," "peasants," "hillbillies" and "yokels" of rural Tennessee. More significantly, subsequent accounts of the trial linked the anti-evolution mindset to the rise of the Ku Klux Klan in the South. Whether or not that is sustainable as history is not the point. The point is that a great and painful wound in American society, which in its essence was not about science and religion but rather about the governance of the United States, the legitimacy of slavery and the social place of African Americans, continued to fester and was given new depth and pain by the science/religion, or if you like evolution/ Bible, debate.

This is why I say that, though of course the issues have been important elsewhere in the world, Americans seems to have had a particularly hard time of it, and that state of affairs continues to this day. After all, as is often pointed out, many leading conservative Christian theologians, in America as elsewhere, saw nothing particularly threatening in the discoveries of the nineteenth century. As with many other things, a path of wisdom that might have been available at an earlier stage was not taken, and instead the war between the two cultures—as in certain literal wars—continued, with attitudes on both sides hardening.

My point is that the present American context, which reflects these culture wars in newer forms, makes these issues much harder for Americans to deal with than they are for the rest of us. But when in the United Kingdom scientist-theologians like Alister McGrath or John Polkinghorne so clearly model ways of thinking in which the two worlds are wisely and richly integrated, most of us nonscientists are quite happy to continue that line of thought and see no need to trumpet our allegiances or to explain our conversions to new ways of thinking. These are not major cultural issues for us. They do not carry—as I fear they often do in America—worryingly direct political implications. Clearly the American issues are important. But it may help to reflect on how they are bundled with larger issues, gaining a lot of their apparent heat from those larger problems rather than from their own innate difficulties. And I say all this not simply in the sense of "we don't see it quite like that." I say it because there is a danger of Americans assuming that everyone else *ought to* share their problems. There is an old story about a gang of youths in Belfast stopping an Indian gentleman on the street. "Are you a Catholic or a Protestant?" they demand. "I'm a Hindu!" he answers. "Okay," they say, "but are you a Catholic Hindu or a Protestant Hindu?" So, am I a fundamentalist creationist or an atheistic scientist? Answer: I'm a Brit.

18

Personal Evolution

Reconciling Evolutionary Science and Christianity

Justin L. Barrett

Justin L. Barrett *is Thrive Professor of Developmental Science at Fuller Theological Seminary and research associate of the University of Oxford's School of Anthropology, where he formerly taught. He lives in California with his wife, and counts his two adult children as his greatest accomplishments in developmental science.*

■ · ■ · ■

POOR MR. BRYSON. I WAS THAT KID—that perfect storm of science geek and Bible reader from a fundamentalist background. I was the top student in Mr. Bryson's high school zoology class. I remember him pulling me aside when the frogs arrived for dissection. "Barrett," he said, "make sure you get one of these double-pithed ones." Then came our big classroom presentations. Mr. Bryson had to watch as his star student argued that evolutionary theory was not so settled; that it was *merely* a theory and that it had some fundamental flaws. I knew—or thought I did—that evolution was a direct challenge to Christianity and I knew my loyalties. Mr. Bryson didn't push back

during any of the other presentations, but he did gently challenge me. I was persuasive to the rest of the class and that made
him uncomfortable.

Fascinated by the living world, I eagerly read what I could about
biology, particularly taxonomy and anatomy. I dissected rattle-
snakes and proudly displayed the extracted organs to babysitters.
I made models of organs out of homemade modeling clay. I recall
a cross-sectioned kidney with which I was particular pleased. I
loved the living world and its wonders, and I wanted to become a
biologist up until my first year of college, when I discovered psychological science and changed course. I also loved God and ac-
quired a deep respect for biblical authority from my parents.
Once I used biblical texts to calculate the age of the earth. It never
occurred to me that Scripture and science could be in conflict. I
loved science, but knew it was constantly changing and thus could
get things wrong. I also knew that our understanding of what the
Bible teaches is imperfect. My background was not one of sophisticated Bible reading, but it was attentive. My childhood home did
not consider that Adam and Eve might not be historical figures,
but it did notice that Adam and Eve and their brood were not the
only people on earth. Who was Cain afraid would kill him after
he was cast out for murdering his brother? If there were other
people coexisting with Adam and Eve, then who were they and
where did they come from? Presumably God specially created
many humans and not just a pair, but was that right?

Science was fascinating and carried authority. In the world in
which I grew up, however, evolution was not science: it was ideology and propaganda. A song I learned in the children's
program at one church my family attended included the lines:
"I'm no kin to the monkey / The monkey's no kin to me / I don't
know about your ancestors but mine didn't swing from a tree."
By the time I made it to college zoology class, however, I was

ready to hear a different message: maybe God used evolution to bring about humans, and maybe that possibility did not fundamentally conflict with either biblical authority or basic Christian theology. I was not yet convinced by evolutionary science, but I was willing to give it a fair hearing.

I cannot say when I finally decided that evolution by natural selection is currently the strongest reading of the evidence. I know it was a gradual process that was complete by the time I ended graduate studies in experimental psychology at Cornell University and began teaching at Calvin College. I wonder how much of my general acceptance of some form of God-endowed evolution was aided by being asked to read and respond to Richard Dawkins's *The Blind Watchmaker* by one of my dissertation committee members at Cornell. I am not suggesting that I found Dawkins's argument against theism convincing. On the contrary, by that time I had received enough exposure to the philosophy of religion to be able to distinguish good arguments from clever rhetoric. But perhaps it was just this security that enabled me to try evolutionary science on for size. If a spirited attempt to defeat God through evolution failed, then maybe evolution was more house cat than hungry lion.

Conspiracy theories about scientists piecing together ordinary bits of bone to make dinosaurs or relying on faulty radiocarbon dating techniques to argue that the earth was hundreds of millions of years old became increasingly absurd once I got to know science and scientists firsthand. The degree to which they are willing to publicly fight each other over seemingly trivial matters, the way jealousy and competitiveness leads to savage personal integrity attacks and the fierce independence of many scientists all make widespread conspiracies very unlikely indeed. Even if the scientific community wanted to manipulate the evidence to cause problems for people of faith, they could not pull it off. And then there is the fact that *they* is *we*: a much greater proportion of God-fearing, Bible-believing Christians make up the ranks of professional scientists than is often supposed. My atheist professors and colleagues did not always make my transition to accepting evolution by natural selection easy. I bristled instinctively at the scorn they heaped on "ignorant fundamentalists" and I saw that evolutionary theory was, for many of them, more a matter of ideology than scientific inference. Smugness is unattractive in Christians and atheists alike.

My training as a psychological scientist also exposed me to the ways psychological dynamics can make evolution difficult to understand and believe—and not just for Christians. When teaching evolution, it is helpful to be sensitive to some basic facts:

- Deep time is hard to understand.

- The idea that one kind of animal can descend into another is wildly counterintuitive and sounds implausible, even if it is true.

- It is much easier to think about the natural world and its processes—including evolution—as having direction and purpose. Human minds are attracted to purpose and meaning.

Furthermore, when Christians are challenged to consider evolutionary science, they are often not just being asked to consider the facts but also to consider their allegiances. Whose team are they on? These Christians may have very good reasons for their worldview and very good reasons for thinking it may be threatened by evolution. Patience and understanding is required.

I am grateful for my long journey toward reconciling Christianity and evolutionary science. It has taught me that Christians who are uncomfortable with any version of evolution—even evolutionary creationism—are not necessarily unintelligent, naive or obstinate. It has also taught me to be suspicious and careful of what ancillary assumptions are being smuggled in with proclamations about evolution. For many people, antitheists and conservative Christians alike, evolutionary science assumes a directionless, godless cosmos. Once we jettison those assumptions, evolutionary sciences take their right and proper place as tools at the Christian's disposal for a clearer theological understanding of humanity, creation and the Creator.

19

The Evolution of an Evolutionary Creationist

Denis O. Lamoureux

Denis O. Lamoureux *is associate professor of science and religion at St. Joseph's College in the University of Alberta. He holds three earned doctoral degrees—dentistry, evangelical theology and evolutionary biology. Lamoureux was a contributor to* Four Views on the Historical Adam *(Zondervan, 2013).*

■ ▪ ■ ▪ ■

EVOLUTIONARY CREATION IS AN EVANGELICAL Christian approach to evolution. It embraces a complementary relationship between biblical faith and evolutionary science. Evolutionary creationists believe that the Father, the Son and the Holy Spirit created the universe and life, including humans, through an ordained, sustained and intelligent design–reflecting evolutionary process.

This Christian view of origins asserts that evolution is teleological. It is a planned and purpose-driven natural process headed toward a final goal—the creation of men and women. Evolutionary creation contends that humans evolved from prehuman ancestors, and that the image of God and human sin were gradually and mysteriously manifested. Most importantly, evolutionary creationists enjoy a personal relationship with Jesus.

It took about twenty years for me to become an evolutionary creationist. Like most people, when I began to think seriously about origins I was trapped in the origins dichotomy. This either/or way of thinking forced me to assume that there were only two credible views of origins: *either* evolution *or* creation.

The destructive power of the origins dichotomy first appeared when I was a freshman biology student at a public university. It took only one introductory course on evolutionary biology for me to step away from my Christian faith. By my senior year I was an atheist. But my story is not unique—many have lost their faith over evolution. It is quite understandable that many churches are worried about their young people studying biology in secular universities.

By God's grace and in answer to my mother's prayers, I returned to Christianity a few years after graduating from university. Through reading the Gospel of John, the Holy Spirit convicted me of my sinful lifestyle. The Lord also revealed to me his unconditional love and amazing forgiveness. Christian faith brought me amazing joy, hope and purpose.

However, the wretched origins dichotomy soon reappeared for the second time in my life. I began attending a wonderful church— but most of its members were still trapped in the either/or thinking of the origins dichotomy. They despised evolution and claimed that it was Satan's weapon to attack university students. In Sunday school I was introduced to creation science (or young-earth creationism). It was presented as *the* Christian view of origins. My teachers claimed that there was a significant amount of scientific evidence to prove that the world was created in just six days about six thousand years ago. They also told me that the godly way to read the six-day creation described in Genesis 1 was to read it literally. I was warned about theistic evolution (what I would later call evolutionary creation) and got the impression that so-called Christians who held this position weren't really committed to Jesus because they didn't fully trust the Bible.

Attending Sunday school completely convinced me that true Christians were young-earth creationists. It wasn't long before I was totally consumed by creation science. I went to seminars and summer schools on young-earth creationism, attended evolution versus creation debates, and became friends with one of the leading creation scientists in the nation. I even promoted my anti-evolutionary views publicly in the journal *Creation Science Dialogue* and ended an article by saying, "I challenge anyone who takes pride in their objectivity to entertain seriously scientific creationism. It may very well be the most important study of your life."[1]

How committed was I to young-earth creation? I walked out of first-year medical school at the University of Toronto and decided to become a creation scientist. I believed that the most important thing I could do with my life was to attack Satan's evolutionists in universities. To equip myself for the battle I decided to pursue a PhD in evangelical theology followed by a PhD in evolutionary biology.

My theological education began at Regent College in Vancouver, British Columbia. My area of specialization was Genesis 1–11. As my diary records on the first day of school, I had an agenda. "The Grand Plan: To declare pure and absolute hell on the 'theory' of evolution." But my Grand Plan was soon challenged, as I discovered what every theology student experiences—interpreting the Bible is more complicated than we were taught in Sunday school.

In a class taught by J. I. Packer, one of the world's most important evangelical theologians, I was told that the biblical creation accounts "were obviously written in picture language." I had read many of his helpful books and knew people who had become Christians through his famous book *Knowing God*. But Packer's claim that the accounts of creation had "picture language" really shocked me, because like most evangelicals Christians I assumed that reading Genesis 1–2 literally was the godly way. And I wasn't alone. After the lecture about fifty of the seventy or so students stormed to the front of the class to challenge Dr. Packer. Needless to say, it was a pretty rancorous exchange.

Rancor also marred a course on the relationship between science and Christianity. Most of the students were young-earth creationists and we merciless attacked our professor, Dr. Loren Wilkinson. One day after class, I cornered him in a narrow hallway and I asked directly what he thought about young-earth creationism. He responded bluntly, "It is error." I still remember how the word "error" shook me.

In the final moments of the course Dr. Wilkinson looked at me hard and said, "Denis, I have a serious concern. If you should ever give up your belief in young-earth creationism, would you also give up your faith in Jesus?" Ouch! I mumbled and stumbled, and didn't really answer. But it was a valuable question, because deep in my heart I knew that my relationship with Jesus was more important than any view on how the Lord had created the world.

The reason my classmates and I were so troubled by the views of Drs. Packer and Wilkinson was that these professors were challenging a massive assumption we held about Scripture. We assumed that God had revealed some modern scientific facts in the Bible. This is known as "concordism" or "scientific concordism." But as I proceeded through my theological education it became abundantly clear that the statements about the physical world in Scripture reflect an ancient understanding of nature. In other words, the Bible features ancient science.

Let me offer an example from Genesis 1, regarding the creation of the heavens. On the second day of creation, God creates a firmament to separate the waters and create a heavenly sea overhead. On the fourth day of creation the Creator puts the sun, moon and stars in the firmament. From an ancient phenomeno-logical perspective, this is exactly what the structure of the world looks like. The blue of the sky gives the impression that there is a body of water above, upheld by a firm structure. The sun, moon and stars seem to be placed in the firmament just in front of the heavenly sea. In fact, the three-tier universe depicted in figure 1 was the best science of the day in the ancient Near Eastern world, and it appears throughout Scripture (for more details, see my book *I Love Jesus and I Accept Evolution*, especially pages 43-70).

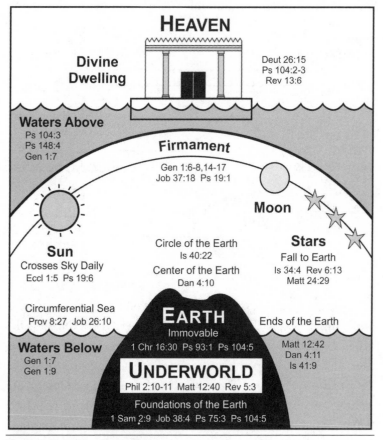

Figure 1. The three-tier universe

Another feature of Genesis 1 that challenged my scientific concordism was its ancient poetry. As figure 2 reveals, the creation account of Genesis 1 is structured on a pair of parallel panels. In the first three days of creation God outlines the boundaries of the universe, and in last three days he fills the world with heavenly bodies and living creatures. Parallels emerge between the panels. For example, God's creation of light on day one aligns with the placement of the sun, moon and stars in the firmament on day four. Evidence like this *within* Scripture led me to conclude that the Bible was not a book of science.

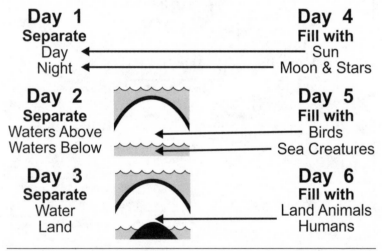

Figure 2. Genesis 1 parallel panels

My theology professors also introduced me to a new way of understanding how God had inspired the writers of the Bible. In order to reveal himself to ancient people, God came down to their level and used their understanding of nature. In other words, the Holy Spirit *accommodated* them by employing ancient science as a vessel to transport the inerrant spiritual truth that the God of the Bible was the Creator of the entire world.

After my PhD in evangelical theology, I entered a PhD program specializing in the evolution of teeth and jaws. Though I now recognized that scientific concordism was not a feature of the Bible, I still believed that evolution was complete nonsense and that there was no proof whatsoever to support it. My plan was to quietly collect scientific evidence that disproved evolution. Once I had completed my degree, I would write a book to destroy evolution and then begin my career attacking evolutionists in universities.

For three and a half years I tried my best to interpret the scientific evidence through various anti-evolutionary theories of origins, but eventually I came to the conclusion held by nearly everyone who has studied evolutionary biology—the scientific evidence for evolution is *overwhelming*. There is no debate within the scientific community regarding evolution, and there are no competing theories. In fact, a 2009 Pew survey of scientists reveals that 97 percent of them accept that living organisms, including humans, were created through an evolutionary process.[2]

In Sunday school I had been taught that there were no transitional fossils in the geological record, but early in my evolutionary education I began to see countless numbers of these fossils indicating that evolution was true. To be sure, finding out that transitional fossils existed was quite shocking at first. But I could not deny this scientific evidence. Let me offer some examples that convinced me.

Reptiles evolved into mammals through a series of transitional animals called "mammal-like reptiles." The fact that this term is used is evidence that transitional fossils exist. Figure 3 shows a couple of examples in the evolution from the simple cone-shaped teeth of reptiles to the complex specialized teeth of mammals. Reptilian teeth function well for grasping and killing animals, but they are not useful for chewing. In contrast, mammalian back teeth have sharp edges that cut up their prey and draw out of more vital nutrients.

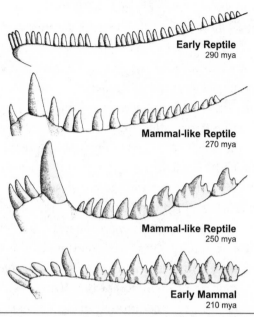

Figure 3. Reptile-to-mammal tooth evolution. Mya=millions of years ago. Redrawn by Braden Barr and based on Robert L. Carroll, *Vertebrate Paleontology and Evolution* (New York: W. H. Freeman and Company, 1988), 196, 365, 386, 406, 408.

Figure 4 presents some transitional fossils in the evolution of jaws from reptiles to mammals. Notably, the bones of the reptilian jaw joint are entirely different from those in mammals. The question arises: How did these jaws evolve? An animal needs a functioning jaw joint to eat and live. Mammal-like reptiles provide the answer. Some mammal-like reptiles had two jaw joints! They had the original reptilian jaw joint and the newly evolved mammalian one. This mammal-like reptile with double jaw joints is proof that transitional fossils really exist.

Alongside my research on evolution, I also studied embryology. It was amazing to discover the incredibly complex set of well-coordinated chemical reactions involved during the development of different creatures. This science filled my soul with awe and strengthened my faith that living organisms reflect the intelligent design of a Creator. As Psalm 19:1 states, "The heavens declare the glory of God." I would add that embryological development from a fertilized egg to a mature adult also declares "the work of his hands!"

In my studying both evolution and embryology, the Lord introduced me to a very important analogy between these two sciences. As a Christian, I believe that God created us in our mother's womb by using *his* natural process of embryology. Christians do not believe that our Creator comes out of heaven to miraculously attach an entire leg or arm to our developing bodies. This is also

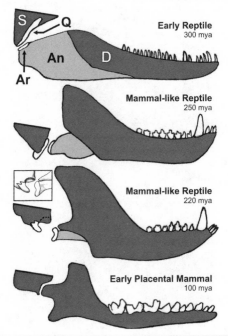

Figure 4. Reptile-to-mammal jaw evolution. An=angular bone, Ar=articular, D=dentary, Q=quadrate, S=squamosal, mya=millions of years ago. Redrawn by Andrea Dmytrash and based on Carroll, *Vertebrate Paleontology*, 366, 382, 390; bottom jaw based on Kenneth D. Rose, *The Beginning of the Age of Mammals* (Baltimore: John Hopkins University Press, 2006), 92.

the case with evolution. Instead of descending from heaven and miraculously placing different creatures on earth, God created all living organisms using *his* natural process of evolution.

Finding this analogy between God's creative action in both embryology and evolution completely freed me from being afraid of evolution. It became evident that science was the study of God's creation and all the natural processes that he had created, including the process of evolution. I began to see that far from being an enemy of Christianity, science is actually a gift from God that declares his glory and reveals how he made the universe and life, including us.

20

Learning from the Stars

Laura Truax

Laura Truax *is the senior pastor of LaSalle Street Church in Chicago. LaSalle has a long history of engaging Christian faith with cultural and societal issues. Truax is currently writing a book on living generously (Eerdmans, 2016).*

> *If you try to stop the people, I tell you the*
> *stones themselves will cry out in praise!*

LUKE 19:40

> *The heavens are telling the glory of God! And all creation*
> *is shouting for joy. Come dance in the forests, come play*
> *in the field. And sing! Sing to the Glory of our God!*

MARTY HAUGEN

■ ▪ ■ ▪ ■

EVEN THE LIGHT BREEZE FAILED TO LIFT the heavy humid air. The summer night gave the mosquitoes free rein on our bare arms and legs as the five of us kids, along with our mom and dad, lay on the grass staring up at the inky sky. This was to be the largest meteor shower in the past several years. My dad was not about to let us to miss it, no matter how hot or buggy the night was. He would hear no complaints or register any concern. "This is central Florida in July! You, lucky duck, are living in paradise."

"Stop talking now and pay attention! There! Right there at the ten o'clock position! See it? *See them?*" My younger brother and I quit squirming and began staring intently at the night sky.

Did we see them? We couldn't have missed them. Stars escaping the black blanket above us appeared to be running for their lives in their downward descent. It was magical.

We stayed for hours—waiting until the cosmic chase was over before we walked back across the field, dragging our old sheets and pillows to our house. For days to follow my little brother would throw sand in the air, reenacting the night of the falling stars.

My father, an atheist, attempted to explain what we had seen. He was a good explainer. As always, my siblings and I tried to pay attention and learn. It was from my dad that I first learned about old stars and new stars, and why some of the stars appear to be winking at us while others stare down at us constant and unwavering. Reading from his *Scientific American* magazine (and dumbing it down significantly), Dad told me about the new theory of "dark matter"—how the universe was expanding and how our earliest human ancestors had looked and walked differently from me.

As I grew up most of my dad's explanations faded. Despite his best efforts, I just didn't have the kind of mind to hold the dense details together. But the wonder, the awe-inspiring majesty, of what I felt in the field that night took root. There was something profoundly noble and true in human curiosity and the pursuit of inquiry. There was something glorious and cosmically magical all around us. Of that much I was sure.

The psalmist wrote that "The heavens are telling the glory of God. . . . Day to day pours forth speech, and night to night declares knowledge" (Psalm 19:1-2 NRSV). There is a freedom of praise that seems to arise unbidden and unforced from God's people throughout the Hebrew texts—a spontaneous gasp of worship when viewing the sky or looking at the mountains. The natural world—one that they understood only in a very rudimentary way—was a place of unfolding wonder and delight. The world around them, *reality itself,* was a revelation of the Almighty God.

Engaging with reality—with *what is*—was one of the marks of a good king during the time of the Israelite monarchy. Whether that reality was the treatment of widows or the threat of foreign powers—there was a lot to be lost if the national leaders did not accurately name the reality of their situation.

When I became a Christian at an outdoor tent revival I was struck by how Jesus had entered a world *I knew.* He used flowers and trees, wind and water to illuminate his Father's truth. He compared my academic stress to birds looking for food and my ache for high school popularity was held up to the luminous grace of lilies growing effortlessly in the fields. Jesus knew the life that was all around me. Jesus didn't come to deal with theoretical problems; he came into the reality of my world. As countless souls before me have said, the incarnation of Christ gave even greater weight to the fact that God revealed himself *in this world of time and space.*

As I have followed Jesus these past three decades, these twin convictions have shaped how I understand God and science. First, I am convinced that the world of our creative God is vast and glorious. Second, I am certain that the Judeo-Christian faith must robustly engage with *what is.*

I was in college when the idea of a young earth began to make its way through my church community. Some of my friends excitedly announced that the young-earth model proved that the growing evidence of evolutionary and cosmic dating couldn't possibly be true. From the beginning I thought that this Christian response disagreed with the reality of what science was learning.

Not only that, but the young-earth argument didn't seem to align with the ever-expansiveness I had experienced with God. As I read the arguments that the earth must be only several thousand years old because of the familial generations described in Genesis, or because of one interpretative lens of Genesis 1–11, I felt less wonder, not more—less awe of our Creator, not a greater sense of glorious mystery. Perhaps most importantly, I felt a growing unease in the hubris of humans who dare to consign the Almighty to one rather rigid understanding of God's Word—a written word which, for the vast majority of Christianity, was interpreted and lived in connection with the real life and the ongoing pursuit of inquiry all around them.

St. Augustine said, "A person who is a good and a true Christian should realize that truth belongs to his Lord, wherever it is found."[1] The faithful Christian need not hide from exploration and the pursuit of inquiry. Surely, I thought, the Creator is delighted when his people are curious, engaged and delighted by the wonder of the universe around us. The Lord pronounced this world "very good," and throughout the ages men and women have asked questions, sought knowledge and been humbled by the unfolding richness of what is around us.

As the worlds of chemistry, physics, biology and the rest have expanded, it seems more important than ever that Christians find a way to incorporate the new knowledge and accepted scientific truths into the lexicon of our worship and wonder of the Lord.

And what a rich time this is, as our understanding of the origins of human life and even the origin and dating of the universe itself become clearer to us through the lens of science! We are now able to see the vastness of space in new ways through the images of the Hubble telescope. The wondrous filigree of the nebula and the cosmic worlds beyond our galaxy excite people around the world to become amateur astronomers.

Again and again when people come face to face with the wonder and glory of God's work revealed in nature there is a rush of excitement, a flush of new understanding and a deep sense of unsatisfied satisfaction as they realize that even with this new kernel of knowledge, they still stand at the very edge of deeper mysteries and longings. I believe that this sense of humble inquiry is what God seeks in all of us.

Jesus said that if the praise of the people was silenced, the very rocks themselves would cry out in praise to God. This feels like what is happening now. The wonder of the universe is on display, and the long and complex story of human origins is becoming clearer through our growing awareness of our genetic code and evolutionary beginnings. It should be a time when all creation—humans included—shout exclamations of wonder and joy as we engage with what is unfolding.

How invigorating it would be if the excitement now exploding in the scientific community was also found on the lips of God's people, as advancing knowledge became another moment of praise and worship.

Perhaps the church might realize that the dichotomy between faith and science is a false one, and that much of scientific advancement is only another expression of God's mysteries being given language and understanding. These advancements don't detract from the awe or the wonder of the living God. Instead, they offer us today what they offered the psalmist earlier: a way to understand the speech poured forth through the days and unveiled through the nights. This understanding ultimately brings us to the place all of our knowledge brings us: to the worship of the Creator, who delights in all that he has made. To God be the glory.

21

So, Do You Believe in Evolution?

Rodney J. Scott

Rodney J. Scott *teaches biology at Wheaton College, where he has worked for over twenty-five years. He is a fellow of the American Scientific Affiliation, and in 2012 he was a Fulbright Scholar in Costa Rica. Rod and his wife, Donna, have two adult children, Janeen and Phillip. Rod and Donna enjoy traveling and spending time outdoors. They attend Church of the Savior Anglican church.*

■ ▪ ■ ▪ ■

"SO, DO YOU BELIEVE IN EVOLUTION?" I get this question all the time. I get it from my students, from the parents of my students, from my friends and from my colleagues. I'm tempted to respond (and to be honest, sometimes I have), "No, I believe in Jesus and I don't believe in Santa Claus, but I don't even ask myself whether I 'believe' in evolution—that's just not a sensible question." For me, the phrase "believe in" is very similar to the phrase "have faith in," and I don't think of evolutionary theory as something that one can or should "have faith in." However, what my glib response misses is the very real fact that some people have had their faith shaken by *assumptions about* evolutionary theory. So in my better moments I try not to be glib, and I try to really explain how I have come to reconcile my beliefs as a Christian with my understanding of evolutionary theory. Where I usually end up (after a long dialogue) is with two simple (but I think profound) propositions. First, God did it. And second, I really don't know how.

How have I arrived at such a position of enlightenment? And what do I really "believe" about evolution? To answer these questions I need to provide a brief sketch of my faith journey and my intellectual journey, which have generally run on parallel tracks throughout my life.

The first time I remember probing a matter of science-faith integration was when I was about six years old. I asked my mom, "If Adam and Eve were the only people God made, who did Adam and Eve's sons marry? Not their sisters, right? Yuck!" My mom couldn't answer my question, and I still haven't answered it to my satisfaction. However, I survived this intellectual crisis as a youngster by applying my two previously mentioned propositions. God certainly was the one who created the human race, but apparently having the faith to believe that didn't also require having all the answers. After all, my mom seemed to believe the biblical story even though she couldn't explain everything.

The next significant step in my faith journey happened while I was in college. Although I considered myself a Christian, my faith really didn't impact the way I lived, let alone how I thought about intellectual matters. Then two things happened that forever changed my approach to faith. First, a traveling evangelist appeared on campus. I don't remember his name or his denomination, but he asked me a question that got me thinking: "If you died tonight, do you know with certainty where you would spend eternity?" For various reasons, my life was in a state that made me receptive to that question. He also directed another question to me and my friend Tom—both of us biology majors: "How can you believe in that evolution stuff? Can't you see that God's the Creator, and not some random process called evolution?"

I remember this question for two reasons. First, even though the evolutionary explanation made sense to me, I didn't let this apparent conflict prevent me from taking a step of faith. Second, I perceived that this conflict did make it harder for my friend Tom to take the same step of faith. Tom was a hardcore biologist and his reaction was very negative. The way he saw it, to become a Christian meant that he would have to "turn off his brain"—and when it came to biology, there was no way that he could do that. I lost touch with Tom after college, so I don't know how things ended up for him, but it causes me great sadness to think that he may have missed out on life in Christ because he was confronted with what I believe was an unnecessary choice.

The second thing that shaped my faith during college was that I was befriended by a group of students from InterVarsity Christian Fellowship. These folks were Christians and they definitely didn't keep their religion separate from the rest of their lives. They actually read the Bible and tried to apply it to their lives. And instead of asking me how I could believe in "that evolution stuff," they welcomed me into their lives and showed me love. The traveling evangelist got me thinking about my eternal destination, and these folks modeled a way that I wanted to live my life here and now. The result was that I dedicated my life to God.

When I entered graduate school I continued to nurture my new commitment to God by attending a "Bible church," meeting with other Christians in small groups and reading books and articles about Christian faith. It was a time of wonderful spiritual growth for me, but also a time of some spiritual and intellectual confusion. When it came to understanding how my chosen profession, biology, related to my faith, I got a lot of mixed and negative messages. Thankfully God provided Chuck, a mentor who helped me to understand some of the complexities of relating science and faith. Chuck taught me that both science and theology are human endeavors, and therefore both are flawed—and this can lead to apparent conflicts. But he also explained that the subjects of those two disciplines, the created order and the Scriptures, are both works of God and should therefore ultimately be compatible.

After graduate school I needed a job. I applied for positions at various institutions and I received invitations for four interviews. One was at Wheaton College. Though I knew nothing of Wheaton during my early years, I had come to hear quite a bit about it since entering the world of evangelicalism. It was (so I was told) "the Harvard of the Christian Colleges." Though I was rather intimidated about the interview, and I was fairly sure that I didn't have the necessary credentials, I ended up getting the job. When I look back on my interview this seems to have occurred *despite* the lack of sophistication in some of my responses. My response to one question specifically stands out in my mind. An administrator asked me how I accounted for the many physiological similarities between humans and other primates. My response was based on something that I'd read in a tract about evolution. I asked this administrator if he knew that the creator of the Volkswagen beetle and the Porsche roadster were one and the same person (Mr. Porsche of course) and that he had used components from the Volkswagen to make the first Porsche. The administrator didn't know that bit of trivia, but he saw the connection between my story and his question, and this bit of hand-waving was enough to get me off the hot seat. (This illustration was actually more compelling in those days, but as I'll mention below, insights from the genome projects make it less compelling now.) From that day on, I've continued to learn and teach, and then learn some more, about biology and about the connections between science and faith.

A defining moment in my career at Wheaton came early in my tenure when I thought I would lose my job because of what I "believed" (or rather, because of what I was uncertain of) regarding evolution. That moment came when a new president, Duane Litfin, took the helm of Wheaton College. Dr. Litfin took his new responsibilities seriously and he wanted to ensure that Wheaton was still holding Scripture in a place of high authority. Among other things, he wanted to know what the Wheaton science faculty believed about evolution. After a series of discussions with the science faculty, Dr. Litfin concluded that there was a range of possible options regarding beliefs about human origins and that some of these were not compatible with Wheaton's statement of faith. At one end of the spectrum (highly compatible with the statement of faith) was a view that completely rejected evolution as having played any role in human origins. At the other extreme (incompatible with the statement of faith) was a view that rejected that the biblical description of human origins is in any way factual. I found myself somewhere in the middle. I honestly couldn't say that I knew with certainty how God created humans, and I was worried that my view would also be deemed incompatible with the statement of faith. My tenuous situation continued for some time, but Dr. Litfin finally clarified his position and allowed for the possibility that a faculty member could be uncertain about exactly how God did it.

These events at Wheaton occurred in the early 1990s, and in today's world the 1990s are ancient history—especially with regard to debates about human origins. Two developments have occurred since then that have dramatically changed the landscape. First, the "genome projects" have highlighted the incredible genetic similarity that exists between humans and other organisms.

Second, the rise of the New Atheists (led by authors like Richard Dawkins, Daniel Dennett and Sam Harris) has revived the notion that science and faith are at war, and that science is winning.

The primary result of the genome projects is that it is not as easy to explain away the similarities between humans and other organisms as it once was. For example, it is no longer simply a question of physiological similarities between humans and chimps, but also a question of striking genetic sequence identities. My old analogy of using parts from one car to build another becomes meaningless now that we know that at least 95 percent of the (genetic) parts are identical. With such a huge number of identical parts, the situation seems more similar to a car maker bringing out a new model of last year's version than making two separate cars. Furthermore, many of the genetic similarities occur in genes that no longer appear to function because of mutations, and which have been interpreted as relics of evolution by mainstream science.

The impact of the New Atheists is revealed most poignantly to me in a few recent conversations I've had with students. These students say that they've lost their faith because they can no longer accept both the findings of science (which they regard as true) and the claims of religion. Whether these students cite the influence of individuals like Dawkins or not (and some do), clearly the arguments of such writers have become part of the cultural milieu in which they find themselves. While these writers are persuasive, what many readers fail to see is that they misuse the authority of science (the study of the *natural world*) to claim that belief in the *supernatural* is irrational. I feel the same way about these students that I feel about my college friend Tom. It breaks my heart to think that they have been confronted with a false choice and have abandoned their faith as a result.

So where does all this leave me, and what, after all, do I believe about evolution? My primary response is to repeat my two basic propositions: God did it, and I really don't know how. Though these two phrases are brief and simple, they summarize several key ideas that shape my faith and my thinking. The first is the idea of faith—God did it. I believe this proposition not because science has demonstrated it, but because I have faith. What I see in the world around me can lead me to faith, but no finding of science can ever demonstrate to me that my faith is justified. That is exactly why it's called faith.

A second idea that undergirds my two propositions is that theology and science are two different but equally valid ways of understanding the world. As human enterprises both sometimes lead to incorrect conclusions and apparent conflicts. However, since both theology (the study of Scripture) and science (the study of creation) focus on different works of God, the ultimate answers of both should be compatible.

Finally, the two propositions suggest specific roles for the church in its interaction with science, particularly regarding the theory of evolution. The primary role of the church is to glorify God. One significant way to do this is to recognize and recount what he has done in creation. Another role of the church is to care for God's creation, which requires an understanding of it. Both of these roles suggest that Christians should embrace the sciences as a way of doing God's will. A third role of the church is to reconcile people to God—and though it's less obvious, an embracing of science is needed here too.

While I have known of many people who have been driven away from the church by controversies related to evolution, I do not personally know of anyone who has embraced Christianity primarily because they were persuaded to reject evolution. While evolutionary theory can be used to support an atheistic explanation for how the world came to be, it does not require that perspective. Why should the rejection of evolutionary theory be a litmus test for Christian faith? It is time for Christians to agree about what we believe *about* evolution—not that we would agree about how God did it (that will never happen!)—but we should agree that evolution is an acceptable option.

22

The Spirit of an Evolving Creation

Surmisings of a Pentecostal Theologian

Amos Yong

Amos Yong *is director of the Center of Missiological Research and professor of theology and mission at Fuller Theological Seminary. He is the author or editor of over three dozen books. He lives in Southern California with his wife, three children and one grandchild, and attends The Bridge, a Foursquare start-up in Pasadena.*

■ · ■ · ■

AS A PENTECOSTAL CHRISTIAN FAMILIAR with miracles, exorcisms, healings, signs and wonders, and spiritual gifts, I have long been concerned with developing a theological framework that can account for Pentecostal-charismatic spirituality and for the world of modern science and technology, the benefits of which are also enjoyed by myself and my community of faith. The most challenging scientific theory, of course, is biological evolution. Within my church circles, there is still widespread belief that evolution is incompatible with Christian faith as a whole, not to mention a Pentecostal-charismatic way of life. There are various strands of issues that I have had to disentangle to open up my reconsideration of this issue.

First, there is the rhetorical dimension. On the one hand, conservative Christians talk about "warfare" between faith and science. On the other hand, there is a long history of naysayers (including a whole group of "new atheists") who presume that all the big questions of life and even religion can be explained by science. There are many reasons why this warfare thesis has persisted; in large part, it is because the naysayers confuse the data and interpretation of science with their own materialistic and naturalistic philosophical presuppositions. I came to this realization fairly early in my graduate studies, and it was the first step that invited me to examine the scientific evidence for evolution.

Second, I have also come to see that what laypeople call "evolution" really is an extremely contested notion among whole groups of scientists. This does not mean that there is not a solid emerging consensus about the main lines of evolutionary development (about which I'll say more in a moment), including the common ancestry of all life forms, but there are many questions regarding the mechanisms of evolutionary change that remain debated. Science, of course, is driven by the quest to answer questions. What religious folk and theologians like me need to do, then, is to pay attention to scientists as they pursue these matters, explore theoretical frameworks for their research and deliberate about this or that proposed hypothesis. Along the way we should confront those atheists (new or otherwise) who would impose their own philosophical interpretations on the scientific data, even as we should be cautious about demanding that scientists conform their research to our own theological commitments.

But what does science tell us? While fully acknowledging that I have not much more than a high school science education (with a couple of college courses following), let me say that I find the following three lines of evidence convincing enough for me to support the work of mainstream science and its evolutionary hypothesis.

(1) The evidence for the age of the earth and the cosmos is undeniable. The light of stars from the farthest reaches of the cosmos travels toward us at the speed of 186,280 miles per second, or 670 million miles an hour. Combined with measurements of the distances between galaxies, there is no question that we can see back as far as 12 billion years ago, suggesting that the universe itself is at least that old. The established rates of decay of isotopes of certain chemical elements have shown that some meteorites and other terrestrial samples are about 4.5 billion years old. We assume the earth is about that old too. In addition, geological studies of sedimentary layers, plate tectonics, continental shifts and fossil deposits provide additional confirmation for an ancient earth.

I have not found the few young-earth perspectives that respond to the scientific data convincing. God could have made the earth and the universe itself as a whole about 10,000 years ago, or even two minutes ago (why not?), all with the appearance of age (including our memories of events before two minutes ago!), but I find indefensible the young-earth theological apologies for what amounts to divine deception.

(2) The second set of evidences concern the geographical distribution of life's diversity. Here I think that even if the fossil record leaves much to be desired (laypeople untrained in the sciences, like myself, are unimpressed by the arguments explaining why we do not find more transitional forms), the geographical isolation of lakes, islands and continental drift makes coherent our understanding that different life forms, including plants, bacteria and animal species, have flourished in different parts of the world. Their development is consistent with the

time frames required by an old earth. More importantly, the patterns of life's expansion traceable through careful study of biological diversity in different ecological, climatological and geographical niches all make sense when situated within an evolutionary framework.

(3) Most recently, and perhaps most importantly for thinking about the thesis of common ancestry involving human beings, the results of the Human Genome Project seem to support evolution. Not only do human beings share 95 to 99 percent of their genetic sequence with chimpanzees, their closest living relatives, but we know enough about how genetic information changes between generations to be able to follow the lineage of many genes through eons of time. Yet to say that humans share the vast majority of their gene structure with chimps, for instance, is not to say that human beings evolved from monkeys. Geneticists and other scientists would correct such a view: shared genetic endowments only point to the conclusion that humans and chimps share a common extinct ancestor. Yet the remarkable similarities of gene sequences again make sense within the evolutionary hypothesis. The evidence pointing to various forms of prehominid groups and their possible relationships to one another is also consistent and coherent across a number of disciplines, ranging from genetic to cognitive and anthropological sciences.

When so much data from such a wide swath of disciplines— from astronomy, geology, zoology and paleontology on the one hand to neurobiology, cultural and linguistic anthropology, and social-psychology on the other hand—are combined, the theory of evolution broadly understood by mainstream science is reasonable to me. Again, this is not to say that we have gathered all the data or that we have it all figured out. No—that is the point of scientific research: to continue to test the theory. But until another theory emerges that explains everything explained by evolution, and explains some things that evolution cannot explain, I am hard pressed to provide reasons why I would embrace the advances in many other arenas of the modern and especially applied sciences—in engineering, medicine, transportation, information technologies and so on—while rejecting the consensus of mainstream science regarding the age of the earth and biological evolution.

But as a Christian who is committed to a high view of Scripture, I still have to understand how the Bible is compatible with, if not complementary to, what science tells us. In particular, I urge that we rely on the work of biblical scholars to help us understand that the ancient narrative of Genesis ought to be read in its historical and cultural context rather than as a (modern) scientific account. I am now convinced that the Scriptures affirm only *that* God created the world, while science fills in the details of *how* that happened. Surely, again, this does not mean that no difficult questions remain. For instance, how should we think about Adam and Eve in relation to evolutionary theory? I look forward to developments in the field of biblical scholarship that will increase our understanding.

As a Pentecostal theologian, finally, part of my scholarship over the last ten years and more has been devoted to understanding what it means to lead a Spirit-filled life in a scientific and even evolutionary world. I am not saying that I have all of the answers—which is in part why I am motivated to engage in dialogue, research and reflection with many other of my colleagues in Pentecostal and charismatic colleges and universities who are also pondering these important matters. But I now see no irredeemably irreconcilable conflict between affirming God's role in guiding the process of the universe and the earth's evolutionary history while working to save the fallen world from sin through Jesus Christ and the Holy Spirit. The Spirit of God continues to accomplish miraculous and redemptive works for those of us who have the eyes of faith, in anticipation of the reign of God to come. Thus I can both pray and expect God to heal and simultaneously prepare to undergo surgery—the Holy Spirit can heal directly or through a surgeon.

The God who has overseen the evolution of life is the same God who visits us personally in Christ and the Spirit. I have met many other evangelicals and Pentecostals along the way who also do not see any irresolvable problems with this attitude, and they also have bolstered my faith in the plausibility and usefulness of such a posture. Yet the conversation is barely begun in Pentecostal circles, and I eagerly look forward to much that is yet to come.

23

Two Books + Two Eyes =
Four Necessities for Christian Witness

Richard Dahlstrom

Richard Dahlstrom *(stepbystepjourney.com) is an author* (Colors of Hope), *conference speaker, Bible teacher and senior pastor of Bethany Community Church in Seattle. In his free time he likes skiing, climbing and hiking with his wife in the Cascade Mountains, "where the Creator speaks so clearly!"*

■ · ■ · ■

I WAS TALKING WITH SOME PEOPLE in my church recently about a significant moment that occurred when their daughter came home from college. Raised in Christian schools her whole life, she'd taken a science course at her Christian university where she encountered, for the first time, a thoughtful and reasoned explanation of evolution—and how, far from being contradictory to the biblical accounts of creation, there were powerful markers pointing toward harmony.

As internally coherent as all this might have been in a vacuum, the experience was devastating to her. She met with her parents one afternoon and explained what she was learning, all of which was in contradiction to her young earth, anti-evolution up-bringing. Silence hung over the coffee until she asked, "What else did the church lie to me about?"

Her parents are thankful she was willing to have the conversation. Among eighteen to thirty-year-olds—the most rapidly declining demographic in the American church—the contradiction between faith and science is one of the main reasons for departure from the faith. It's as if the church has created a Y in the road: intellectual integrity one way; faith the other. Thousands stand at this crossroads that the church has unwittingly created and walk away from their faith. The greatest tragedy of this departure is that the Y in the road is a fabrication of religionists, not a construct of either God or the Bible.

How the church has come to this point is beyond the scope of this essay—I'm not writing as a church historian. I'm also not writing as a scientist, so the particular details of DNA, evolution and the geological evidence for the age of the earth are also beyond this essay. I write as a pastor of a church in a vibrant city filled with university students who are studying biology, physics, medicine, archeology, astronomy and every other scientific discipline known to humanity. These students are the future, and we who are called to be the presence of Christ for them should make sure we're following Jesus in our ministries—not the religious leaders whom Jesus confronted on a regular basis to expose the truth that they "tie up heavy, cumbersome loads and put them on other people's shoulders" (Matthew 23:4). In contrast, Jesus' burden is light, and his followers thought it was important to "lay . . . no greater burden than these essentials" (Acts 15:28 NASB). We must learn, both from the example of Jesus and the teachings of the early church, that it's eminently easy to demand things of people "in God's name" as a prerequisite for acceptance by God—things which Christ himself never demanded. When we do this and people walk away, they're not rejecting the gospel. They're rejecting a caricature, and we will face judgment for it. When we invite people to Christ, whether as pastors, teachers, youth workers or parents, we need to be careful to invite people to Christ, not to a religious system of our own fabrication. We can avoid building false walls by embracing two central principles regarding how we perceive the gospel.

1. God Has Spoken Through Two Books

Psalms 19 and 104, Romans 1 and 10, the book of Job and the parables of Jesus all make it clear that God has spoken not only through the Scriptures, but also through creation. God has spoken so clearly that all people are "without excuse," because the evidence of God's character is present for all to see.

In particular, Paul declares that "God's invisible qualities—his eternal power and divine nature—have been clearly seen, being understood from what has been made, so that people are without excuse" (Romans 1:20).

We believe that these words from Romans 1 are true for all people in all times—including scientists in the twenty-first century. When the people who study stars with telescopes and the people who study tiny molecules with microscopes reach independent conclusions that both point in a similar direction regarding origins, we'd do well to listen. If we don't, we'll be forced to create our own subculture of alternative science, one that swims upstream against not one discipline, but virtually every field of science—from astronomy to geology, and chemistry to biology.

I was a lay consumer of this alternative science for decades, parroting evidence I'd heard which pointed to a very young earth: there wasn't much dust on the moon, the speed of light was slowing down, the geological strata might have happened quickly via punctuated catastrophes, there are no transitional forms, and so on.

As a pastor whose undergraduate work was in music and architecture, I was ill-equipped either to confirm or deny these declarations. But confirm them I did, because they were offered up "in Jesus' name," and came with an implicit understanding that all faithful Christ-followers believe these things about origins.

"Of course they're right," I'd think to myself, because the truth is that this was the only view I'd heard. The insidious thing about subcultures is that they're entirely self-referential. When we sit in a closed circle and speak only with people whose thoughts and beliefs mirror our own, we become convinced that our views are truth.

Years later, after moving to the city, I encountered thoughtful Christ-followers who believed in the risen Jesus—and in evolution. Their reasons for belief were the same in both cases: overwhelming evidence! These new friends helped me see that the precise conditions necessary for life to exist (something called "the fine tuning of the universe") provide markers that point to something more than randomness. As Freeman Dyson, former physics professor at the Institute for Advanced Study at Princeton University and one of the most brilliant and interesting astrophysicists living today, said, "The more I examine the universe and the details of its architecture, the more evidence I find that the universe in some sense must have known we were coming."[1] This is just one example of how the book of creation is pointing us to the Creator.

2. Two Eyes: Humility and Interdependence Are in Keeping with Jesus' Character

As compelling as the scientific evidence for evolution is, questions and objections quickly arise because of apparent contradictions with the Genesis narrative. "Evolution requires death, and there was no death until Adam and Eve sinned." "Humans were made directly from dust"—these are just two of many reasons people who love God and his Word have a hard time embracing evolutionary creation. Schooled in what's called the literal, or holistic, interpretive method, they believe that the plain literal meaning of any text is the best choice for interpreting the Bible.

None of those who agree with that method apply it to every word of the Bible. We don't believe the sun is literally rising, for example, in spite of that plain meaning. Scientific discoveries forced a rethinking of the plain and literal reading. Eventually, the church caught up with the rest of the world and agreed that the earth orbits around the sun and rotates on its own axis.

This is an example of the ongoing challenge we face as readers of both the Bible and the book of creation. What's needed is a sense of humility when reading both books. With respect to the Bible, humility is important because history shows us how easy it is repeatedly to get our interpretations wrong—including the justification of slavery, colonialism and genocide.

Jesus himself is scathing in his assessment of just how wrong the religious leaders of his day interpreted the Bible. "You study the Scriptures diligently because you think that in them you have eternal life," he says (John 5:39), but goes on to say that they're unwilling to come to Jesus that they might actually have life. For them the Scriptures had become a sort of rule book or legal code. Their wooden interpretations of it were so distorted that in the end they conspired to kill the very Messiah who was the object of their longing. In other words, it's easy to get things wrong, and it's for this reason that humility is in order.

Among other things, humility means embracing the reality that we don't know everything. St. Augustine, in a book called *The Literal Meaning of Genesis*, wrote about the dangers of pastors and theologians pontificating beyond the scope of their authority:

Usually, even a non-Christian knows something about the earth, the heavens . . . and this knowledge he holds to as being certain from reason and experience. Now, it is a disgraceful and dangerous thing for an infidel to hear a Christian, presumably giving the meaning of Holy Scripture, talking nonsense on these topics; and we should take all means to prevent such an embarrassing situation, in which people show up vast ignorance in a Christian and laugh it to scorn. . . . If they find a Christian mistaken in a field which they themselves know well and hear him maintaining his foolish opinions about our books, how are they going to believe those books in matters concerning the resurrection of the dead, the hope of eternal life, and the kingdom of heaven, when they think their pages are full of falsehoods and on facts which they themselves have learnt from experience and the light of reason?[2]

Natural science and hermeneutics (the science of interpreting the Bible) are both best practiced with humility and interdependence, allowing each to be informed by the other. Where this is present, our mutual understanding and appreciation of each other's disciplines leads to greater clarity in our own disciplines. Where this is lacking, one ends up entrenched in either a spiritless materialism, or a fundamentalism constantly at odds with the findings of the physical sciences. In a world where God has told us that he speaks through both the Bible and the book of creation, neither of these options is palatable.

On a recent flight I found myself seated next to an astronomy professor. Though I'm not astronomer I do visit NASA's "Astronomy Picture of the Day" website as a regular part of my morning devotions to remind myself of the vastness of our universe. Thus I had questions for the professor, and we talked for most of the flight about the expansion of the universe, the Big Bang, and other marvelous mysteries of space. He was an older man but still immensely curious, telling me that the more he learned, the more aware he was of how much he didn't know. As we began our descent, he asked me what I did for a living. When I told him I was a pastor, his eyes lit up and he said, "We need each other, you and I!" He went on to talk about how theology had answers to questions he thought science would never be able to uncover, and vice versa. "Two books," I said, as I heartily agreed and explained what the Bible had to say about creation pointing to God.

It was a good and hopeful conversation, and we need more of them in our world. For that to happen, we need to believe that the books of creation and the Bible aren't in contradiction, and we need to allow them to inform one another. Then our curiosity and creativity—which are ours because we're made in God's image—can be released to allow people to discover, create, heal and much more "in Jesus' name." Such a paradigm will enable joyful Christ-followers to be wholly engaged in being the presence of hope in our glorious—yet broken—world, instead of retreating into narrow subcultures where the false dichotomy between faith and reason becomes a wall that is, for too many, insurmountable.

24

Finding Rest in Christ, Not in Easy Answers

Kathryn Applegate

Kathryn Applegate *is program director at BioLogos. She received her PhD in computational cell biology from The Scripps Research Institute in La Jolla, California. A native Texan, she now lives in Michigan with her husband and two small children, where they are active members of a Presbyterian (PCA) church.*

■ ∙ ■ ∙ ■

I WAS SO HAPPY, NESTLED ON A COUCH between two good friends at our church's annual women's retreat. The speaker was describing a biblical approach to counseling. In particular, she emphasized our gifts as women for helping and nurturing. "We are life givers," she said. "This goes all the way back to the Garden. The world wants to remove distinctions between men and women, and to remove our status as image bearers." *Amen,* I thought. "Our story is that of creation," she went on. "What's the *world's* story for how we got here?" "Evolution!" a chorus of voices replied. I sighed and made eye contact with another biologist in the group. She looked like I felt—disappointed.

I still love that church, which my husband and I attended until we moved away. I love the high view of Scripture and worship that the leaders maintain. I love the emphasis on living in community and serving our world. I am grateful that even though my view on origins is different from that of my former pastor, he and the rest of the elders supported me over the years as I sought to reconcile science and theology. Yet as my women's retreat experience shows, it is not always easy to be a biologist in a Bible-believing church. I think it would even be harder to *become* a Christian if you were a biologist confronted with such an environment.

I haven't always accepted evolution. In fact, for a long time I didn't want it to be true. All I knew was that it seemed to be a creation story for non-Christians. It's true that many use evolutionary science to support an atheistic worldview. But I've come to understand that sliding into atheism is by no means a necessary outcome of understanding the science.

I grew up deep in the heart of Texas, which is deep in the heart of the Bible Belt. As a kid, I loved studying nature and could usually be found climbing trees, catching butterflies or looking at all manner of things under my grandfather's old microscope. I also loved God and the Bible. I came to faith in Christ at summer camp when I was nine and spent much time memorizing Scripture and attending church with my family. I even repeatedly shared the gospel with my dog, Tuffy, because Jesus surely died for her sins, too!

As a teenager I thought often about matters of science and faith. Once I presented my dad with a list of Big Questions ranging from "What's on the other side of a black hole?" to "What activities are there in heaven?" But when I went off to college, I wasn't prepared for the onslaught of new ideas that would challenge my faith. Centenary College, in Shreveport, Louisiana, is a true gem of the South—a small liberal arts college with Methodist roots. It was a rich intellectual environment, and there were many opportunities for fellowship with other Christians.

But like many college students, I found it all too easy to avoid serious commitment to any congregation or campus ministry group, and this "untethering" made me feel lonely and isolated in my faith.

In my freshman year I took a Bible survey class and learned about other ancient creation stories that were eerily similar to the early chapters of Genesis. Also, there were lots of other writings that weren't included in the New Testament; the early church had decided which books to include. And there were loads of textual inconsistencies I'd never noticed before. How could I have never heard about these things? It had never occurred to me to ask *how* the Bible had come to be and how we could still say this is *God's* Word when there was clearly so much human intervention. With no resources to work through these apparent problems, the very foundation of my faith seemed to crumble. Meanwhile, many of the Christians I knew seemed either anti-intellectual or hypocritical—myself included. I continued to read my Bible every morning and go to church most weeks (often sitting in the back row, analyzing everything I heard with paralyzing skepticism), but I had serious doubts about Christianity.

I set out to major in biophysics and eventually added a math major as well. I was good at science and loved the elegance and beauty of it, but I avoided the one area I felt might drive a final nail in the coffin of my weak faith—evolution. One biology professor on campus was an open advocate of Darwin and evolution. I am ashamed to say that it occurred to me more than once that she might be the Antichrist.

In the last semester of my senior year, the Lord gave me a gift. I began to attend a Bible study at the home of a young lecturer in the English department. She and I met regularly over coffee that semester. I grilled her with four years of unanswered questions. She was the first person I met who displayed both a deep love for Christ and a passion for scholarship and rigorous thinking. Through her, God worked powerfully in my life to restore my faith.

With a voracious appetite for theological and scientific knowledge, I arrived in California for graduate school at The Scripps Research Institute. I joined a computational cell biology laboratory to study the dynamics of the cell's internal scaffold, the cytoskeleton.

While my work was not directly focused on evolution, the topic was everywhere around me, just as young-earth creationism was part of the culture where I grew up. Confronted almost daily by my ignorance of the science of evolution, I decided to embark on a study. By this point, I was confident that if Christianity were true, it could withstand even the toughest questions. And if it couldn't, I didn't want any part of it.

So I set out to read a series of books from multiple perspectives. In 2006, world-famous geneticist Francis Collins published *The Language of God*, in which he described his radical conversion from atheism to Christianity and laid out a number of compelling lines of evidence for evolution. My heart filled with joy as I turned the pages and saw so many things falling into place. It didn't answer all my questions (and the more I learn, the more questions I have), but I found a place of rest, both mentally and spiritually. This was another gift from God.

In the coming months I connected with others who shared Collins's evolutionary creation perspective. They loved science because it revealed the secret workings of God's good creation and helped them worship God better. But they also loved the Bible and didn't try to gloss over the difficult parts. And most importantly, they loved Christ's church, despite what her members were saying about science. I felt like I had come home.

Several years later, I still have lots of questions. Some are about science and some are about theology and biblical interpretation.

An example of a science question I have is the extent to which evolution is driven by natural selection acting upon random mutation. This is a hotly contested question in evolutionary biology. This might confuse many Christians, since we so often hear scientists say that evolution is a fact. What is *not* contested is that all the diversity of life of earth, including humans, shares a common ancestry. To be sure, many relationships have yet to be worked out in detail, but it seems undeniable, especially given the many kinds of genetic evidence uncovered over the last couple of decades, that humans share a common biological ancestor with other species. However, the specific processes driving evolution are still under investigation.

The common ancestry of humans with other species raises immediate questions about biblical interpretation. Doesn't the acceptance of evolution require the rejection of the Genesis account of human origins? Indeed, it *is* hard to reconcile evolution with the traditional understanding of Adam and Eve as the first (and only) parents of the human race. But evolutionary science is silent on whether Adam and Eve were historical figures; it merely states that there was never a time when just two people walked the earth. Perhaps Adam and Eve were the first two people with whom God began a covenant relationship. In the fullness of time, he called them out for a purpose, just as he did with Abraham, Moses, David, Elijah and pretty much everyone else in the Bible. God made a covenant with Adam and Eve, which they broke when they fell into sin. As our representatives, their sin became our sin.

I am aware that many Old Testament scholars, even conservative ones, feel there are good reasons to think Adam and Eve were *not* historical figures. I can respect that. I don't think the gospel hangs in the balance either way. Christianity depends on the historical life, death and resurrection of Jesus, whose sacrifice on the cross redeemed us from our sins.

I also have questions about deep time and suffering. Humans have been around about two hundred thousand years, which sounds like a long time until you consider that the universe is 13.8 *billion* years old. What was God doing all that time before we existed? I don't know, but I suspect he was delighting in his good creation. And what exactly does Paul mean in Romans 8 when he says that the creation was subjected to futility and groans for the revealing of the children of God? It is hard for me to attribute all pain, death and suffering to humanity's fallen state (as many Christians do) for two reasons. First, as the fossil record makes clear, physical death has been around since the dawn of life on earth, long before humans existed. Part of our creation mandate—given before sin entered the picture—was to subdue the earth (Genesis 1:28). The need for subduing implies that some amount of natural disorder existed in the beginning, as part of God's plan. Second, while death is surely our enemy to be vanquished in the end, the Bible often describes death and suffering being used to serve redemptive ends. Jesus declared that a man wasn't born blind because of anyone's sin, but so that "the works of God might be displayed in him" (John 9:3). Paul says we must rejoice in our suffering, because suffering produces endurance, character and hope (Romans 5:3-4). Indeed, "our light and momentary troubles are achieving for us an eternal glory that far outweighs them all" (2 Corinthians 4:17). And most importantly, the cross of Christ was not some cosmic cleanup job, a Plan B devised when humans first sinned. It was ordained from the beginning (Acts 2:23). We may not ever fully understand the depths of these mysteries, but it is clear that "all death is the result of sin" is too easy an answer.

Evolution is an elegant and beautiful means by which the earth brings forth living creatures (Genesis 1:24) under God's providential hand. The idea of sharing an ancestor with a chimpanzee doesn't offend me in the slightest. Rather I am humbled when I consider that God chose humans from all his creatures to bear his image and become his sons and daughters. I look forward to the day when Christians will worship the awesome God of creation without rejecting the testimony of his works revealed in the fabric of the created order.

- 25 -

Safe Spaces

Richard J. Mouw

Richard J. Mouw *is professor of faith and public life at Fuller Theo-logical Seminary, where he served as president for twenty years. Before coming to Fuller he taught philosophy at Calvin College. In 2007, Princeton Theological Seminary awarded him the Abraham Kuyper Prize for Excellence in Reformed Theology and Public Life.*

■ ▪ ■ ▪ ■

I WENT TO A LARGE PUBLIC HIGH SCHOOL in New Jersey. Somebody gave me a book to prepare me for the challenges that I would face taking biology courses and the like. It was published in 1938 by Eerdmans Publishing Company, and it was called *Monkey Mileage from Amoeba to Man*. I don't remember much about it, but I do remember one little poem that we were encouraged to learn in a fundamentalist Bible group that I belonged to in that high school. The poem went like this:

> Once I was a monkey, long and thin,
> Then I was a froggy with my tail tucked in.
> Then I was a baboon in a tropical tree,
> And now I am a professor with a PhD.

As you might imagine, that did not get me very far in wrestling with the issues.

Between ages fourteen and sixteen I worked summers at a fundamentalist Bible camp, where there was an older student who was already studying at Wheaton College. He discovered a book by Bernard Ramm called *A Christian View of Science and the Scriptures*, published in 1954. Ramm was one of the great leaders of the post–World War II neoevangelical movement. In that book Ramm discussed, in a way that I had never heard anyone discuss before, the problems with believing in a universal flood. As an evangelical, he set forth two possible alternatives to young-earth creationism. One was theistic evolution (what many now call evolutionary creation), and the other was what he called progressive creationism, which he preferred. He wanted to view God as intervening at various points with specific miracles—for example, investing the evolving human person with a soul and so on. This friend of mine read the book the way other kids our age were reading *Playboy*. We kept it in a paper bag and sneaked off into the woods to read passages out loud to each other. This book and that experience forever formed my own understanding of the issues.

Much later, when I was a professor at Calvin College, I invited Bernard Ramm to the college to speak. We spent several hours together, and after telling him how much the book had influenced me, I asked him—since this book was very controversial, and there were many attacks and many people questioning whether he was a real Christian—"Do you ever regret writing that book?"

"Oh, no," he said. I'll never forget what he said next: "All the criticism I ever received was worth it, just to know that there would never be a student of mine who, after studying with me, went off to Harvard and lost his or her faith because I never allowed them to wrestle with the kinds of issues that I raised in that book."

We face a younger generation that's wrestling with these issues, and we need to wrestle with them ourselves. But I'm also convinced that it's a larger task. It's a task for the scientists and other scholars in the Christian community. It's a task for the presidents and deans and others at Christian colleges and seminaries. It's a question for pastors. How do we deal with the spirituality issues here? At the heart of it, there are some spiritual challenges that are presented to us. It's tough to wrestle with them, and that's why we need a safe space. It's tough for all kinds of pastoral and academic leaders. These are also tough questions for theological reasons. I still haven't settled on a plausible answer to the historical Adam question, for instance. I want to hang on to what the apostle Paul says: that it's by one person that sin came into the world, and it's by one person that we have been rescued from that sinful condition. I'm struggling with it, but I need safe places in which to explore with other Christians who are also willing to explore.

I can say from personal experience that it's tough for people in academic leadership. I had a case of my own where a wealthy donor was very upset with a professor at Fuller Seminary who had written a piece criticizing some aspect of the intelligent design (ID) movement. This person was supporting Fuller Seminary and also some of the leaders in the ID movement. The faculty member who had written the piece came up for tenure, and the wealthy donor called me and said, "I know you have the right to veto it even though the faculty has approved it. If you allow this professor to receive tenure, I'll never give another penny to Fuller Seminary." Usually when somebody says that, they probably have given a penny at some point, but not much more than that. In this case, we had received literally millions of dollars from this donor. I had to say to the person, "I'm sorry, but you need to find another seminary to support if that's the way you feel about it." Those are tough things to say; those are tough decisions to make. But we need to be willing not only to make those decisions, but to try to create a theological and spiritual environment in which we don't have to make those decisions as often as we do.

One of the great Reformation-era confessions is the Belgic Confession. It states, "We know God by two means: First, by the creation, preservation, and government of the universe, which is before our eyes as a most elegant book wherein all creatures great and small are so many characters leading us to contemplate the invisible things of God, namely, God's power and divinity."[1] How do we get evangelical believers to *celebrate* God's creation? How do we get them even to get *excited* about the fact, spiritually, that the earth may be millions and the universe billions of years old, and that human beings have evolved from lower forms of life?

I've got to say, I've never heard a good sermon on the first five days of creation. I've heard a lot of people mention the first five days and say that they were literal days, but the preachers that I've heard seldom get any spiritual or theological joy out of those first five days. I want to celebrate the first five days of creation in that passage in Genesis 1. Anne Lamott has said there are only basically three kinds of prayers: "help," "thanks" and "wow."[2] When it comes to creation, we need a lot more "wow" prayers, and we need a lot more "wow" sermons.

When it comes to spelling out the details of what God cares deeply about—of what God takes loving delight in—it's important to be clear about the fact that it's not all about us, about God's human creatures. That's a rather fundamental thesis that I think is fairly obvious in light of what we read about God in the Bible—

that *God loves the whole creation*. Psalm 104 gives good evidence for this claim. It's a hint at God's creative power and goodness that hardly says anything about human beings: "Praise the LORD, my soul. LORD my God, you are very great; you are clothed with splendor and majesty. The LORD wraps himself in light as with a garment; he stretches out the heavens like a tent" (Psalm 104:1-2). After a while the Psalm gets to goats and cattle and springs of living water, but it barely mentions human beings. Toward the end it talks about the fact that we too go to work every day and do things. But there's very little about human beings. The point is that God takes delight in the nonhuman creation.

The fact of God's sovereign rule over all things and his call for us to acknowledge and implement that rule is very clear in the creation account. A lot happened between God and the creation before human beings ever came on the scene. Even if there were only five days, they were whole days in which God created. He didn't just say, "Let there be" and there was. He contemplated. He took delight in. He called light into being, and then he said, "That's good." I think he took a while thinking about how good that was. Then he separated the sea from dry land and said, "That's good." He created vegetation and said, "That's good." Then he created the sun and moon and said, "That's good."

I especially like the scenario of God creating living creatures. He looked down at puddles and said, "Let those puddles teem with living things. Let there be swarms," and there were. God looked down on those puddles with all those little bugs and fish swimming around, and God says, "That's good." God took delight in that. And finally, when he created humans, the first thing God said to them was, "You better take care of this."

God cares about the nonhuman creation. That's why that wonderful verse we all know and love, John 3:16, says, "For God so loved the world." The word there for "world" is *cosmos*, the created order. "For God so loved the *cosmos* that he gave his one and only Son, that whoever believes in him shall not perish but have eternal life." And then there's that very important next verse, "For God did not send his Son into the *cosmos* to condemn the *cosmos*, but to save the *cosmos* through him." We need to work on this idea.

I've learned quite a bit from my Catholic friends on this issue. One of my dear friends—a blessed memory; she died about a year ago—was Margaret O'Gara, who taught theology at St. Michael's College at the University of Toronto. Margaret and I spent about eight years working on Catholic-evangelical dialogue together at the Institute for Ecumenical and Cultural Research in Collegeville, Minnesota. Margaret really attempted to understand evangelicalism. A couple of us who were evangelicals were sitting around one night at dinner, and Margaret said, "I really admire evangelicalism—there's something in it that really attracts me. But the one thing I can't get is this whole creation thing. Why are you people so hung up on a literal creation?" So we tried to explain to her what's going on with people who care deeply about Genesis's portrayal of God as the Creator of all things. Some evangelicals may get it wrong, we explained. We may go in the wrong direction on this issue sometimes, but there's a deep impulse there that we take seriously. Then Margaret said this, "Don't you evangelicals realize that God is slow?"

It hit me that we have a hard time with that. We want instantaneous conversions. I remember the testimony of a guy at my church when I was a kid. He said, "I was an atheist on August 3, 1948, and I walked into this church, and ten minutes later at this spot"—he would walk to the spot—"it happened. I was a new creature in Jesus Christ." There's something wonderful about that, and we really like it when things happen fast. "Name it and claim it"—that's another part of it. We want instantaneous healing. We have a hard time with a slow God. And yet, God is slow.

In this challenge I was immensely pleased to come across a wonderful paragraph in a 1991 *Christian Scholar's Review* essay by Father Ernan McMullin, who taught in the Notre Dame philosophy department for several decades. Father McMullin was a priest from Ireland who was a very distinguished person in philosophy of science. He affirmed that over millions of years there have been "uncountable species that flourished and vanished [and] have left a trace of themselves in us." The Bible, he says, shows God preparing the world "for the coming of Christ back through Abraham to Adam." Is it too much of a stretch, he asks, "to suggest that natural science now allows us to extend the story indefinitely further back?" And then this wonderful passage follows:

> When Christ took on human nature, the DNA that made him the son of Mary may have linked him to a more ancient heritage stretching far beyond Adam to the shallows of unimaginably ancient seas. And so, in the Incarnation, it would not have been just human nature that was joined to the Divine, but in a less direct but no less real sense all those myriad organisms that had unknowingly over the eons shaped the way for the coming of the human.[3]

I find McMullin's words to be inspiring. It's a theme that adds to our understanding of the creation. That long process beginning in the shallows of unimaginably ancient seas was not wasted time. It was preparation for the one who would come with healing in his wings. For God sent the Son into the cosmos not to condemn the cosmos, but that the cosmos might be saved through him. That healing will only be complete when the Savior returns and announces, "Behold, I make all things new" (Revelation 21:5 KJV). What he will renew in that act of cosmic transformation is all the stuff that he carried in his own DNA and that he took to the cross of Calvary.

We must create safe spaces under the cross for discussion of such tremendously important issues.

Notes

Chapter 1: From Culture Wars to Common Witness

[1]James K. A. Smith, *Letters to a Young Calvinist: An Invitation to the Reformed Tradition* (Grand Rapids: Brazos Press, 2010).

[2]N. T. Wright, *The Challenge of Jesus: Rediscovering Who Jesus Was and Is* (Downers Grove, IL: InterVarsity Press, 1999), 94-95.

[3]Ibid.

Chapter 2: Who's Afraid of Science?

[1]John H. Walton, *The Lost World of Genesis One: Ancient Cosmology and the Origins Debate* (Downers Grove, IL: IVP Academic, 2009).

Chapter 4: Learning to Praise God for His Work in Evolution

[1]E. P. Wigner, "The Unreasonable Effectiveness of Mathematics in the Natural Sciences," *Communications on Pure and Applied Mathematics* 13, no. 1 (February 1960): 1-14.

[2]Carl Sagan, *Cosmos* (New York: Random House, 1980), 193.

Chapter 5: An Old Testament Professor Celebrates Creation

[1]Tremper Longman III, *How to Read Genesis* (Downers Grove, IL: IVP Academic, 2005), 25.

Chapter 6: Embracing the Lord of Life

[1]G. K. Chesterton, *Orthodoxy* (Garden City, NY: Doubleday, 1959), 15.

Chapter 11: Boiling Kettles and Remodeled Apes

[1]Carl Sagan, *Cosmos* (New York: Random House, 1980), 193.

[2]C. S. Lewis, *Mere Christianity* (New York: HarperOne, 2001), 136-37.

[3]Holmes Rolston III, "Genes, Brains, Minds: The Human Complex," in *Soul, Psyche, Brain*, ed. Kelly Bulkeley (London: Palgrave Macmillan, 2005), 23.

[4]Lewis, *Mere Christianity*, 77-78.

Chapter 13: A Scientist's Journey to Reflective Christian Faith

[1]Daniel Taylor, *The Myth of Certainty: The Reflective Christian and the Risk of Commitment* (Waco, TX: Jarrell, 1986), 96.

Chapter 19: The Evolution of an Evolutionary Creationist

[1]Denis Lamoureux, "Philosophy vs. Science," *Creation Science Dialogue* vol. 8, no. 3 (1981): 3.

[2]"Public Praises Science; Scientists Fault Public, Media," Pew Research Center (July 9, 2009), 37. www.people-press.org/files/legacy-pdf/528.pdf.

Chapter 20: Learning from the Stars

[1]Augustine, *On Christian Teaching* 2.18, Oxford World's Classics, trans. R. P. H. Green (Oxford: Oxford University Press, 1997), 47.

Chapter 23: Two Books + Two Eyes = Four Necessities for Christian Witness

[1]Freeman Dyson, *Disturbing the Universe* (New York: Harper & Row, 1979), 250.

[2]Augustine, *The Literal Meaning of Genesis* 1.19.39, Ancient Christian Writers (Mahwah, NJ: Paulist Press, 1982), 42-43.

Chapter 25: Safe Spaces

[1]Belgic Confession, Article 2, in *The Creeds of Christendom, with a History and Critical Notes*, vol. 3, ed. Philip Schaff (Grand Rapids: Baker Books, 1996), 384.

[2]Anne Lamott, *Help, Thanks, Wow: The Three Essential Prayers* (New York: Riverhead Books, 2012).

[3]Ernan McMullin, "Plantinga's Defense of Special Creation," *Christian Scholar's Review* 21, no. 1 (1991): 79.

THE BIOLOGOS FOUNDATION

BioLogos Books on Science and Christianity

BioLogos invites the church and the world to see the harmony between science and biblical faith as they present an evolutionary understanding of God's creation. BioLogos Books on Science and Christianity, a partnership between BioLogos and IVP Academic, aims to advance this mission by publishing a range of titles from scholarly monographs to textbooks to personal stories.

The books in this series will have wide appeal among Christian audiences, from nonspecialists to scholars in the field. While the authors address a range of topics on science and faith, they support the view of evolutionary creation, which sees evolution as our current best scientific description of how God brought about the diversity of life on earth. The series authors are faithful Christians and leading scholars in their fields.

Editorial Board:

- Denis Alexander, emeritus director, The Faraday Institute
- Kathryn Applegate, program director, BioLogos
- Deborah Haarsma, president, BioLogos
- Ross Hastings, associate professor of pastoral theology, Regent College
- Tremper Longman III, Robert H. Gundry Professor of Biblical Studies, Westmont College
- Roseanne Sension, professor of chemistry, University of Michigan
- J. B. Stump (chair), senior editor, BioLogos

www.ivpress.com/academic

biologos.org

Finding the Textbook You Need

The IVP Academic Textbook Selector
is an online tool for instantly finding the IVP books
suitable for over 250 courses across 24 disciplines.

www.ivpress.com/academic/